DISCARD

THE CHICAGO PUBLIC LIBRARY

FORM 19

A YANKEE ACE IN THE RAF

MODERN WAR STUDIES

Theodore A. Wilson *General Editor*

Raymond A. Callahan
J. Garry Clifford
Jacob W. Kipp
Jay Luvaas
Allan R. Millett
Dennis Showalter
Series Editors

A Yankee Ace in the RAF

The World War I Letters of Captain Bogart Rogers

Edited by *John H. Morrow, Jr., & Earl Rogers*

University Press of Kansas

© 1996 by the
University Press of Kansas
All rights reserved
Published by the
University Press of Kansas
(Lawrence, Kansas 66049),
which was organized by the
Kansas Board of Regents
and is operated and funded
by Emporia State University,
Fort Hays State University,
Kansas State University,
Pittsburg State University,
the University of Kansas,
and Wichita State University

Printed in the
United States of America

10 9 8 7 6 5 4 3 2 1

The paper used in this publication meets the minimum requirements of the American National Standard for Permanence of Paper for Printed Library Materials z39.48-1984.

Library of Congress
Cataloging-in-Publication Data
Rogers, Bogart.
A Yankee ace in the RAF : the World War I letters of Captain Bogart Rogers / edited by John H. Morrow, Jr., and Earl Rogers.
p. cm. — (Modern war studies)
Includes index.
ISBN 0-7006-0798-6 (alk. paper)
1. Rogers, Bogart—Correspondence. 2. World War, 1914-1918—Aerial operations, British. 3. World War, 1914-1918—Personal narratives, American. 4. Great Britain. Royal Air Force—Biography. 5. Fighter pilots—United States—Correspondence. I. Morrow, John Howard, 1944-   . II. Rogers, Earl, 1927-   . III. Title. IV. Series.
D602.R54   1996
940.4'4941—dc20   96-7191
                                                CIP

British Library
Cataloguing in Publication Data
is available.

CONTENTS

Preface vii

Introduction 1

1 TORONTO *August–November 1917* 17

2 TEXAS *November 1917–January 1918* 33

3 ATLANTIC CONVOY *January–February 1918* 51

4 CHATTIS HILL *February–March 1918* 62

5 TANGMERE, SCOTLAND, AND FRANCE *April 1918* 80

6 MOUNT KEMMEL *May 1918* 98

7 MONTDIDIER AND NOYON *June 1918* 122

8 SECOND BATTLE OF THE MARNE: CHÂTEAU-THIERRY *July 1918* 138

9 AMIENS AND THE SOMME OFFENSIVE *August 1918* 154

10 CAMBRAI AND THE HINDENBURG LINE *September 1918* 171

11 YPRES AND LYS *October 1918* 194

12 ARMISTICE *November 1918* 211

13 POSTWAR *December 1918–May 1919* 229

Postscript 257

Index 261

PREFACE

Most letters are written by ordinary people. Those collected in this book were written by a young man barely twenty years old when he began, not yet twenty-two when he finished. I was more than twice that age when I first read them, yet the fact of his extreme youth escaped me, even though he was my father. The letters simply showed a talent beyond his years. He wrote them with a spirit of hope and humor and optimism, despite the grim realities of aerial warfare that he tried to impress on readers of *Popular Aviation* some twelve years after the end of World War I:

> Everytime I hear someone speak of the war in the air—the late war and the French air—as a gallant and romantic business, a modern counterpart of the chivalrous strife of old, I break right out laughing.
>
> It was a cold, calculating, deadly occupation—sans chivalry, sans sportsmanship and sans any ethics except that you got the other fellow or he got you. If you could shoot him in the back when he wasn't looking, or bring odds of ten to one against him, so much the better.
>
> The only people who fully understand the war in the air are the fellows who fought it.

Bogart Rogers was one of them. As John Morrow writes in the Introduction, Rogers's letters close a "gap in historical awareness." They also mirror the personality of an American youth who was raised in a talented family and who became in time a very successful writer.

These are letters of the heart and spirit, reflecting lasting values: a war flier's courage, a fighting squadron's bonds of friendship, a flight leader's discipline, a pilot's faith, a sergeant's loyalty—and a young man's love for and courtship of the girl he left behind, and to whom he paid the highest of all compliments by refusing to varnish the truth. All these enduring qualities of the human experience Bogart learned during seven months of violent aerial combat.

Bogart's father, Earl Rogers, was perhaps the most dramatic public defender in Los Angeles history. His showy courtroom style profoundly changed the practice of criminal law. "Earl Rogers was the greatest jury lawyer of his time," said famous Chicago attorney Clarence Darrow, whom Rogers defended for jury tampering.

His mother, Hazel Belle Green Rogers, was a flamboyant beauty, very popular with California's Spanish aristocracy. His sister, Adela Rogers St. Johns, one of the first all-round woman reporters and first woman sports writer, was a rising star in the Hearst newspaper empire.

Bogart was born on June 24, 1897, in Los Angeles. He attended elementary school there until the family moved to San Francisco, where his father was chief defense counsel in the sensational municipal-graft trials that battered San Francisco for three years following the earthquake of 1906.

After the trials, the family returned to Los Angeles. Bogart entered Hollywood High School at a time when Hollywood was mostly farmland, not yet the glamorous movie capital of the world. He played fullback on the varsity football squad and ran the 440 on the track team. In 1915, the graduating seniors elected him class president. In the fall of that year, he enrolled as a freshman at Stanford University and spent the next two years in undergraduate studies preparing to follow in his father's footsteps. But, as these letters reveal, world events would alter his plans.

During the spring of 1917, college campuses buzzed with the talk of war. At Stanford University, a contingent of undergraduates had already volunteered for service in the Ambulance Corps and, with much fanfare, had gone to France at the end of February.

April 1, Palm Sunday, was the beginning of spring break. Stanford students were anxious. President Woodrow Wilson had called a special session of the newly elected Sixty-fifth Congress to convene on Monday. Bogart, then a nineteen-year-old sophomore, was thinking about the Junior Opera.

He and a fraternity brother in Kappa Sigma were writing the lyrics to that year's production, *Pirate for a Day*. Isabelle Young, a petite history major with a lovely soprano voice, was to play the female lead. Bogart was finding excuses to drop around the Kappa Alpha Theta sorority house, where she lived.

They saw little of each other during Easter week, however. Isabelle's father, a Stanford alumnus, and her mother, a native San Franciscan, had arrived from Albany, Oregon. Isabelle joined them, leaving Bogart free to argue with his fraternity brothers about the big question that could alter their lives: Would the United States join Britain and France in the fight against Germany?

On April 6, Good Friday, Congress voted yea, and President Wilson signed the declaration. America was at war.

From the frantic weeks following Easter to summer vacation, Bogart was torn. Should he enlist in this "adventure" or finish his education? In June, he left Stanford before senior week because his mother was ill. When he boarded the train for Los Angeles, Isabelle stood on the station platform waving goodbye, unaware that two long years would pass before they saw each other again.

During the summer of 1917, war fever mounted, with marching and banners and songs. Bogart learned that the Royal Flying Corps was recruiting Americans for training in Canada. By late August, he and four other young men from Los Angeles—Palmer "Pop" Taylor, Bill Leaf, Tom Whitman, and Bill Taylor—had signed with the RFC and departed for Canada.

Pop Taylor died of scarlet fever in the training camp in Texas. Bill Leaf was shot down less than a week before the Armistice. Tom Whitman was hit by friendly fire and fell to pieces in the air. Bill Taylor was shot down in flames.

Bogart was the only one of the five who came home from the war.

In 1993, I obtained a copy of *War in an Open Cockpit: The Wartime Letters of Captain Alvin Andrew Callender*, a book discussed more fully in the Introduction.

Callender is a central figure in some of Rogers's later letters. Photos of his crash and his cross are a haunting witness to his tragic death. A compulsion to track down the editor of *War in an Open Cockpit* led me in April 1993 to a meeting in Pensacola, Florida, with Gordon W. Callender, Jr., nephew of Alvin Callender. It was a thrill to discover a common bond, but more important it was Gordon Callender who urged me to seek a publisher for these letters. I want to thank him.

My mother, Isabelle Young Rogers, who carefully preserved the letters in a cedar chest for more than three-quarters of a century, has read them, at the age of ninety-eight, one last time and found a kind of gentle peace in knowing that there was a reason she kept them all those years. They would become a permanent tribute to the only man she ever loved.

Sacramento, Calif., September 1995　　　　　　　　　　　　　　　　　　E.R.

INTRODUCTION

First Lieutenant Bogart Rogers, a native Californian and Stanford man, arrived on the western front in May 1918 a newly minted fighter pilot posted to the Thirty-second Squadron, Royal Air Force (RAF). He had joined the ranks of young Americans who were so eager to fly for the Allied cause that they volunteered for service in foreign armed forces. The United States had remained neutral while the greatest war in history engulfed Europe like a tidal wave. Rogers, like many others, could not wait to be swept away on the tide; he rushed into it and, ultimately, into the swirling maelstrom of the western front.

Before the entry of the United States into World War I in April 1917, American volunteers flew with the French Air Service in the Lafayette Escadrille, which entered combat in May 1916 above the Battle of Verdun. In 1917 and 1918, the more than 200 American volunteer pilots who formed the Lafayette Flying Corps became the combat leaders of the embryonic American air arm.

The first units of the United States Air Service (USAS) arrived at the front in the spring of 1918 in the Toul sector. They fought over American battlefields—Château-Thierry in July, Saint-Mihiel in September, and finally the Meuse-Argonne toward the end of the war. The greatest aces of the USAS, Captain Edward V. (Eddie) Rickenbacker and Lieutenant Frank Luke, won the Medal of Honor.

The deeds of the fliers of the Lafayette Flying Corps and the USAS are well known. Yet the experiences of Bogart Rogers and the other Americans who flew with the British have not received comparable recognition in the public's awareness of the American aviation experience in World War I. A highly selective review of the literature on these aviators indicates this gap in historical awareness.

PREVIOUS PUBLICATIONS AND PROBLEMS

In 1968, Aaron Norman's substantial book, *The Great Air War*, did not acknowledge that American aviators flying with the Royal Flying Corps (RFC) and then the Royal Air Force had achieved sufficient victories to be ranked among the aces.[1] Norman's list of American aces did include Captain E. W. Springs, who had first flown in the Eighty-fifth Squadron, RAF. Norman noted of the book *War Birds: Diary of an Unknown Aviator* that the unknown aviator was Lieutenant John McGavock Grider, "who flew and died in 1918 with the American Expeditionary Force" and whose diary was "anonymously edited for publication by another American ace, Capt. Elliott White Springs."[2] In fact, Grider had flown in the Eighty-fifth Squadron, RAF, with his best friend Springs, and *War Birds* consisted essentially of Springs's vivid fictionalized recounting of their experiences flying with the British.

War Birds was one of the most famous aviation yarns to come out of the war.[3] Princeton graduate Elliott White Springs and two friends, John McGavock Grider and Laurence Callahan, belonged to a group of 210 American pilots who, en route to Italy, were commandeered by a Royal Flying Corps desperate for new recruits. The three trainees caroused their way through a flight training that claimed the lives of cadets less fortunate, but certainly more sober, than Springs and company.

In a further twist of fate, in the spring of 1918, when the three pilots were about to be assigned to American units in France, they were "kidnapped" by Canadian ace "Billy" Bishop for his new Eighty-fifth Squadron, composed of English, Canadian, Scottish, Australian, New Zealand, South African, and Irish aviators. The Americans were fortunate in their commanders: Bishop's successor was the moody Irishman Edward "Mick" Mannock, Britain's leading ace and squadron leader, who was credited with seventy-three kills before his death from ground fire in late July. A little more than a month before, Grider had fallen in combat. A short time later in June, Springs himself was shot down, his face and mouth painfully lacerated from the butt of his Vickers gun.

After convalesence, Springs became a flight commander in the new 148th Squadron, with which he completed the war as an ace with eleven confirmed victories. The 148th Squadron was one of two American squadrons (the other was the Seventeenth) formed in June 1918 and officially attached to the RAF. The 148th finished the war with sixty-six enemy planes to its credit, only three victories behind Eddie Rickenbacker's Ninety-fourth Squadron, USAS, which had been in action three months longer.

In a visit to the United States in 1919, Edward, Prince of Wales, formally awarded the Distinguished Flying Cross to American aces who had flown with the British, Springs among them. Yet Springs became convinced of the lack of recognition of Americans who had served with the British, as his biographer Burke Davis observed: "He complained frequently, if always in private, that hundreds of U.S. aviators had been neglected because the public had been regaled exclusively with the feats of the Lafayette Escadrille and its American successors on the French front, including Rickenbacker's squadron."[4]

In 1925, Springs began to write about his wartime experiences. *Liberty* magazine serialized "The Diary of an Unknown Aviator" under the title "War Birds" during the latter part of 1926. Its sensational reception led to a book contract and the sale of film rights to Metro-Goldwyn-Mayer. Critics, particularly those in England, deemed it a major contribution to war literature.

Since 1960, a number of articles recounting the experiences of Americans who served in the RFC/RAF have appeared, primarily in the *Cross and Cockade Journal*. One of the very few books on the subject is a small one published in 1978, *War in an Open Cockpit: The Wartime Letters of Captain Alvin Andrew Callender, R.A.F*, edited by Gordon W. Callender, Jr., and Gordon W. Callender, Sr.[5] The letters chronicle the aerial exploits of a young Louisianan who joined the RFC in Canada in the summer of 1917. In May 1918, Callender joined the Thirty-second Squadron, RAF, and was credited with ten victories by the end of October.

On October 30, the RAF registered its greatest success of the war, shooting down sixty-seven Germans and losing twenty-nine aviators killed or missing and eight wounded. The Thirty-second Squadron suffered four of those casualties, among them Callender. Shot through the lungs in combat with Fokkers, he crashed just inside the Allied lines north of Valenciennes and died of his wounds within the hour without ever regaining consciousness.

Callender's letters provide some context for Bogart Rogers's letters, as he joined the Thirty-second Squadron less than two weeks after Rogers. In a letter of August 4, 1918, Callender indicated that the number of Americans equaled the number of Englishmen and was exceeded only by the Canadian contingent: "In the squadron now we have 6 Englishmen, 2 Scotchmen, 1 Irishman, 7 Canadians, 6 Americans, 2 Australians, and 1 South African. Pretty nearly every breed of white man there is, isn't it?"[6] In a letter to his sister dated September 29, Callender observed that "A" Flight had the "best record of Huns of late, . . . 'Bud' Hale, from Syracuse, New York, and Bruce Lawson, from South Africa being particularly good, and California Rodgers is pretty good too."[7]

In addition, Callendar's letters reflect the RAF's attitude toward and method of registering aerial victories. In the letter of August 4, Callender admonished his family to "only quit telling me to be an 'ace,' because we don't have those kind of things in the British Army, except four in a pack of cards."[8] On August 12, he noted:

> Officially, now, I have two Huns crashed and four "out of control." . . . "[O]ut of control" means that clouds or ground mist prevented your seeing it crash on the ground.
>
> These fights that we get are usually 10 to 25 miles over [the lines] escorting day bombers so that you have no way of getting ground confirmation either.[9]

In his last letter, of October 30, Callendar, musing that he never had a fight on his side of the lines, concluded:

> How many Boche planes have I demolished? Well, that is hard to say— the squadron record book shows that I have shot down fourteen, but except for one which was shot to pieces in the air I wouldn't like to swear any of the rest were demolished. Once I shoot a machine out of control I find it doesn't pay to stick around to see what happens after its five or ten minute course to the floor.[10]

The editors of *War in an Open Cockpit* noted the elaborate cross-checks done by the RAF to confirm the combat reports of aviators. Squadron reports went to RAF headquarters in France, where staff officers sought official confirmation of enemy losses using observation posts and squadron reports. Headquarters did not keep individual scores, which were noted in the squadron combat reports that also formed the basis for recommendations for medals, and it refused to recognize the notion of the ace, despite its many star pilots. Its offensive strategy meant that definitive verification of "kills" was often difficult, if not impossible.

The final book in this selective historiography is James J. Hudson's *In Clouds of Glory: American Airmen Who Flew with the British During the Great War*.[11] It contains the combat biographies of twenty-five of the American "aces" who flew with the British and one narrative of a squadron in which a large number of American pilots served. Hudson, himself a fighter ace in World War II, echoed Elliott White Springs: "Unfortunately, those American aces who flew only with the British in World War

I are little known by the American public. It is for this reason that this book has been written."[12]

By Hudson's reckoning, more than 300 Americans served, primarily in 1918, with British flying units. Fifty-one were killed in combat, 32 were wounded, 32 became prisoners of war, and 8 were missing. The 28 American aces who flew with the British scored some 294 victories.[13]

Yet Hudson indicated that it is difficult to ascertain exactly who was American. Even in the case of Bogart Rogers, documents in the Royal Air Force Museum, Hendon, list him as "British" in nationality. Hudson considered this circumstance typical of that of many Americans in the RAF, "since it was sometimes difficult to regain American citizenship after enlisting with the British forces."[14]

Historian Bradley King has noted that the Directorate of History of the National Defence Headquarters, Canada, lists 455 Americans in British squadrons and concludes that "we will probably never know how many there were, as some preferred to remain known as 'Canadian,' while others were known to be Americans only by their immediate colleagues."[15]

It is also difficult to agree on who was an ace and on how many victories the aces won. *Air Aces of the 1914–1918 War* is probably the most comprehensive compilation of the aces of all the combatants in the European war.[16] It lists Captain S. W. Rosevear, with twenty-three victories, and Flight Sub-Lieutenant J. J. Malone, with twenty, as the top-scoring American aces serving with the British. Neither appears in Hudson's book, which instead cites William L. Lambert, with twenty-one victories, and Francis W. Gillet, with twenty. *Air Aces* lists Captain F. L. Hale as the next highest scoring American ace, with eighteen victories. Yet according to Hudson, the records of the Thirty-second Squadron credit Hale with only eight kills.[17] Neither Callender nor Rogers, whom Hudson credits with eight and six victories, respectively, based on the records of the Thirty-second Squadron, appears in *Air Aces*.

Callender's letters make it abundantly clear why such discrepancies have existed and will remain. Ultimately, even the most meticulous researcher must acknowledge the difficulty of ascertaining exact victory tallies for British pilots, because of the nature of the offensive air war that they waged. This circumstance certainly does not detract from the achievements of RAF pilots, but it should give pause to any who claim exactitude in their quantification of victory scores.

The literature conclusively establishes the presence of American pilots in the RAF. How did a substantial number of American aviators come to fill the ranks of British squadrons?[18]

In late August 1914, Major General Sir Sam Hughes, Canadian Minister of Militia and Defense, suggested to Lord Kitchener of the British War Office that Canadian and American aviators and soldiers be recruited into the British armed forces. Kitchener replied that no enlistment was to be undertaken outside Canada.[19]

In December 1914, an English electrical engineer living in New York, Warner H. Peberdy, pointed out to the War Office that Canada and the United States had young men who would qualify as pilots. The War Office and RFC did inquire in February 1915 about enlisting "British-born" aviators, but it was still early in the war.

Active recruiting in Canada for the Royal Naval Air Service (RNAS) and the RFC began in 1916—the RNAS in May, the RFC in August, as the RFC suffered heavy losses in the Battle of the Somme. Facing an acute manpower crisis that could derail the expansion planned for 1917, the War Office and the RFC became convinced of the necessity of a large-scale training program in Canada. Late in December, RFC Canada, which would establish a British organization to train thousands of Canadians for air warfare, took wings under the command of Lieutenant Colonel C. G. Hoare. Cadets would receive preliminary ground and flight instruction in Canada and complete their training in Britain.

The number of recruits in 1917 disappointed Hoare, who began to look to the United States as a likely source of fighter pilots. Only a few Americans volunteered for the RFC and the RNAS through 1916, but so many thousands had joined the Canadian Expeditionary Force that it had formed several exclusively American battalions.

By June 1917, Hoare had concluded that it was necessary to recruit American volunteers, and he decided to do so on American soil. Hoare persuaded Brigadier General G. O. Squier, chief of the United States Army Signal Corps, to allow an RFC recruiting office to be established in New York. Hoare had almost recruited Quentin Roosevelt, son of former American president Teddy Roosevelt, with the father's connivance, until they concluded that the press might raise "a song and dance about an ex-President's son."[20] Quentin Roosevelt joined the United States Air Service and was killed in action over France in 1918.

By late September, half of Hoare's recruits came from the United

States. Beyond more immediate prospects for combat and glory in a tempered and tested air arm, enlistees in the RFC joined primarily because they were "tired of waiting for their own country to take her place with the Allies."[21] Hoare would continue to recruit in New York until State Department pressure forced the closure of the New York office in early February 1918, by which time the recruiting crisis had passed.

Hoare and Squier had cooperated for their mutual benefit. American universities could apply the RFC's training methods in their schools of military aeronautics, while the RFC could escape the Canadian winter at three training bases near Fort Worth, Texas, starting in the fall of 1917.

The first contingent of trainees from the schools around Toronto arrived in Texas before Camp Taliaferro, with its three airfields—Hicks, Everman, and Benbrook—was finished in December 1917. When RAF Canada left Texas in mid-April 1918, it had trained more than 400 pilots for the American services, in addition to its 1500 RAF Canada cadets, most of whom were Canadian.

The training took some four and a half months. After military training came instruction in the airplane, the engine, aeronautics, map reading, and wireless telegraphy. Flight training at Lower Training Squadron entailed two to three hours of dual-control flying, five to six hours of solo flying, and thirty to forty landings. Higher Training Squadron required some thirty hours of instruction in cross-country flight, wireless, photography, observation, and bombing. Gunnery School offered three weeks on aerial tactics and fighting. Finally, the newly minted lieutenant received his wings and was sent to England for additional training to compensate for the inferior performance of the American Curtiss JN-4 "Jenny" biplane used in Canada compared with the Avro 504-J used in England.

Training combined strict discipline with a lenient attitude toward failure in flying examinations. Flying discipline was lax, and accidents were rife. The short dual-control period, designed to move the cadets to solo flight as quickly as possible, probably led to many crashes. Yet "dash" was encouraged, though some students demonstrated it to excess. A cadet recounted that student pilots from Camp Borden at Deseronto, outside Toronto, would hover above Canadian Pacific train tracks until the express passed, and then swoop on the train from behind, bump their wheels on the coaches' roofs, dip low in front of the engine, and turn away when they were so far ahead that no one could see the numbers on their rudder.[22]

In Texas in the winter of 1917/1918, the same cadet noted: "The death rate was out of all proportion. There were no curbs on any type of flying; every good stunt ever heard of was attempted by anyone crazy enough to try it."²³ RFC/RAF Canada lost 129 cadets in flying accidents, one for every 1902 hours logged—actually a better record than those of other RAF training units. To put such losses in perspective, many aspiring pilots in all countries were killed in training accidents. In fact, aviators in all countries were as likely to die in accidents as in combat.

Ultimately, RFC/RAF Canada sent 2500 pilots overseas in numbers increasing from 20 in June 1917 to an average of 230 a month in 1918. Bogart Rogers was among the 240 cadets who graduated in December 1917 and the 247 sent overseas in January 1918. Had the war continued into 1919, RAF Canada would have supplied about one-fifth of the reinforcements needed in Europe. As S. F. Wise has observed:

> From mid-1917 onwards a considerable and steadily increasing stream of badly needed pilots, the bulk of their training completed, had arrived at British ports and, after final training, had been sent to service overseas. Their presence and that of the Americans also trained by RAF Canada was crucial to the winning of the air war."²⁴

AIR WAR OVER THE WESTERN FRONT

These pilots were posted to fight the first air war at its height over the western front.²⁵

During World War I, the airplane evolved from an instrument of reconnaissance used individually in 1914 to a combat weapon in 1918. It played a significant role in the war, first through reconnaissance or artillery observation, and later, deployed en masse, as a weapon for fighting, bombing, and strafing. Air services that had begun the war with some 200 frontline airplanes consequently had 1500 to 3000 combat aircraft at the front in 1918.

The Battles of Verdun and the Somme in 1916 marked the true beginning of aerial warfare, as both sides committed larger air arms to the western front in order to gain aerial superiority. The British and French pursued an offensive aerial policy and overall military strategy, the British unrelentingly and inflexibly, while the Germans husbanded their resources and fought defensively.

Even before the Somme offensive began in July 1916, the RFC, which had 426 pilots on July 1, was seriously concerned about manpower shortages. By mid-November, at the official end of the battle after more than four months of a relentless offensive of infantry contact patrols and of reconnaissance and bombing missions over enemy lines, the RFC had lost

499 aircrew killed, wounded, or missing—more than its total complement at the beginning of the battle.

In 1917, the air war rapidly evolved into a war of attrition with the evolution of massed fighter tactics, bombing raids, and close air support. It was grim and brutal, and the notions of chivalry and sportsmanship, the concept of aerial fighting as a sport, so prevalent in the early RFC, eroded. The RFC carried the fight to the Germans regardless of circumstances.

At least fighter pilots received two superior single-seat craft in 1917— the incredibly maneuverable Sopwith Camel and the Royal Aircraft Factory's SE-5. The SE-5, in its 1918 version the SE-5A, which the Thirty-second Squadron flew, was faster and easier to land than the Camel; its performance at altitude, better; and its inline Hispano Suiza engine, quieter. Its steadiness in a fast dive made it a better gun platform than the Camel, while its speed enabled it to break off combat at will. If it did not have the high-altitude performance and maneuverability of its best opponent in 1918, the vaunted Fokker D-7, its speed and diving ability enabled it to more than hold its own through the end of the war.

In 1918, aerial combat was a grueling and costly mass struggle of attrition. The basic tactical unit—the French escadrille of twelve planes, the German *Fliegerabteilung* of nine, and the British squadron of eighteen— often operated in groups like the German fighter circuses of sixty planes in the attempt to achieve aerial superiority.

When Rogers arrived on the western front in May 1918, the RAF, formed in April, had already surmounted the massive German Ludendorff offensive of late March and was refitting depleted squadrons during the lull in operations. The British army, coordinating its use of tanks, artillery, and airplanes, was preparing to strike at Amiens in early August, a battle that would initiate its inexorable advance to the end of the war. Bogart Rogers's Thirty-second Squadron was one of six fighter squadrons in the RAF's Ninth (HQ) Brigade, and fourteen fighter squadrons overall, assigned to support the attack.

In August, the RAF had more than 1600 combat planes on the western front, 55 percent of which were pursuit planes, rendering it primarily a "fighting arm." It was the only force on the western front so organized, as the Germans, French, and Americans balanced fighter and observation strength. RAF aircrews concentrated on offensive patrols, ground attacks on frontline enemy positions, and bombing attacks against the Somme bridges and occasionally against German airfields.

The Germans, though severely outnumbered, asked and gave no quarter. The RAF lost 847 aircraft in August, 100 of them—the war's highest one-day toll—on the first day of the attack on Amiens on August 8, when it suffered 86 casualties, including 62 killed, missing, or taken prisoner. Only the declining ability of the German air force to replace its losses enabled RAF Headquarters to grant the fighter units some respite from the all-out offensive in early September. But the Germans could still occasionally mount fierce opposition to British aerial incursions. As late as October 30, the day on which Alvin Andrew Callender fell, the British encountered heavy resistance from the German air arm.

From the beginning of 1918 to the Armistice, the RAF suffered total casualties of over 7000 men, more than 3700 in combat. Casualties peaked in September at 1023 (588 in battle and 435 from other causes), and declined to 941 in October. These losses indicate the risk that Bogart Rogers and other fighter pilots faced. More than 50 percent of British pilots trained during the war became casualties. The Royal Air Force traced the careers to October 31, 1918, of nearly 1500 pilots sent to France from July to December 1917. Sixty-four percent were killed, wounded (or sick), or missing; 25 percent had been transferred home; and only 11 percent were still in France. Only about 25 percent of all the pilots completed a tour of duty of nine months, a chastening thought should one be tempted to minimize the toll of flying in World War I.

The air war over the western front was no place for the faint of heart. German authors occasionally referred to the generation of young men who fought the war as the "iron youth." The soldiers from other countries were no less stalwart. The airmen of all countries, volunteers all for a service in which, as an anonymous British infantry sergeant once said, "you only falls once," stood out even in such company as a breed apart, the heroes of the Great War. The letters of Bogart Rogers present the experience of one of those courageous youth.

J. H. M., Jr.
Athens. Ga.
September 1995

NOTES

1. Aaron Norman, *The Great Air War* (New York: Macmillan, 1968), 529-30.
2. Ibid., 137-38.
3. The basis for the following discussion of Springs and *War Birds* is in Burke Davis, *War Bird: The Life and Times of Elliott White Springs* (Chapel Hill: University of North Carolina Press, 1987), 1-131.
4. Ibid., 86.
5. Gordon Callender, Jr., and Gordon Callender, Sr., eds., *War in an Open Cockpit:*

The Wartime Letters of Captain Alvin Andrew Callendar, R.A.F. (West Roxbury, Mass.: World War I Aero, 1978).

6. Ibid., 72.
7. Ibid., 76.
8. Ibid., 72.
9. Ibid., 73.
10. Ibid., 79.
11. James J. Hudson, *In Clouds of Glory: American Airmen Who Flew with the British During the Great War* (Fayetteville: University of Arkansas Press, 1990).
12. Ibid., 220.
13. Ibid., 219.
14. Ibid., 141.
15. Bradley King, "Americans in the Royal Flying Corps: Recruitment and the British Government," *Imperial War Museum Review*, no. 6: 94.
16. Bruce Robertson, ed., *Air Aces of the 1914–1918 War* (Letchworth: Harleyford, 1964).
17. Hudson, *In Clouds of Glory*, 266n.16.
18. For the following discussion on training, unless otherwise noted, see S. F. Wise, *Canadian Airmen and the First World War: The Official History of the Royal Canadian Air Force* (Toronto: University of Toronto Press in cooperation with the Department of National Defence and the Canadian Government Publishing Centre, 1981), 1:26–42, 62, 64, 74–75, 76–107, 117–18.
19. King, "Americans in the Royal Flying Corps," 88.
20. Ibid., 92.
21. Ibid., 94.
22. Wise, *Canadian Airmen and the First World War*, 106n.91.
23. Ibid., 106.
24. Ibid., 118.
25. The material for the discussion of the air war comes from John H. Morrow, Jr., *The Great War in the Air: Military Aviation from 1909 to 1914* (Washington, D.C.: Smithsonian Institution Press, 1993).

A YANKEE ACE IN THE RAF

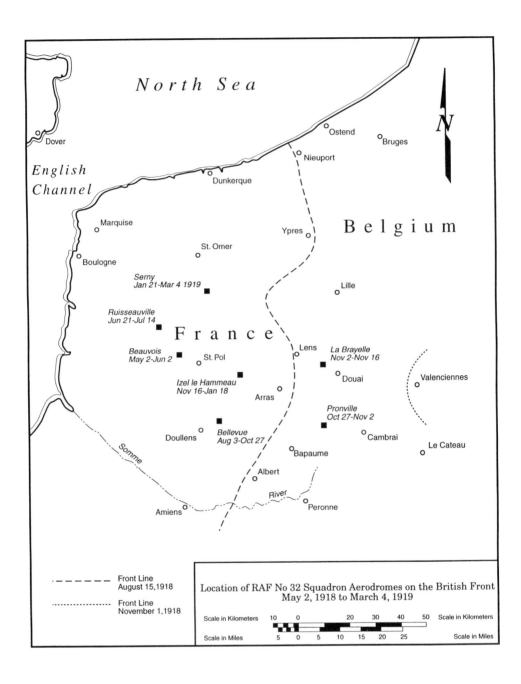

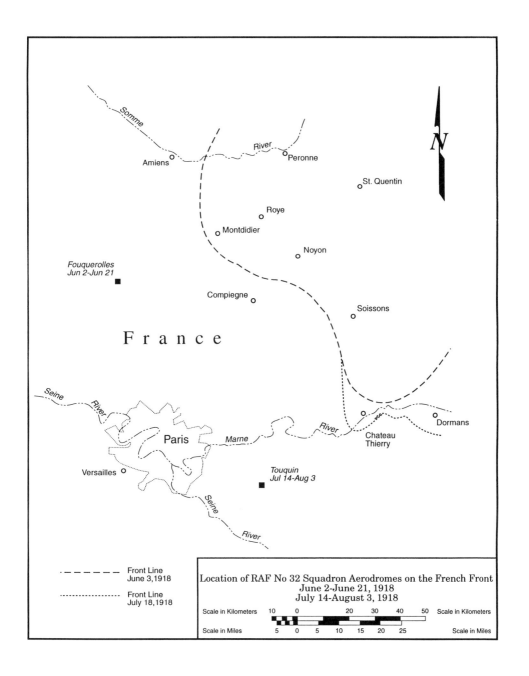

1. TORONTO *August–November 1917*

In March 1917, before the United States entered World War I, the Royal Flying Corps (RFC) established two camps at which British World War I instructors would train pilots in Canada: Camp Borden, 50 miles northwest of Toronto, and Camp Mohawk at Deseronto, 130 miles east of Toronto. In July, the RFC established the No. 4 School of Military Aeronautics at the University of Toronto, which was fashioned after schools in England. It was basically a preflight school for the initiation of newly arrived cadets into the mysteries of military flying and the thrill of marching to pipes and drums.

Bogart Rogers arrived at the No. 4 School of Military Aeronautics in mid-September. By the first week in November, he had completed the six weeks of ground school and was posted to Camp Mohawk for elementary flight training in the JN-4a Curtiss Jenny. He soloed after just under five hours of dual instruction, and a week later he was sent to Camp Borden for cross-country and altitude flying.

. . . .

                                                  Victoria
                                                  Tuesday
                                                  August 28, 1917
                                                  Empress Hotel

Miss Isabelle G. Young
434 W. Sixth Street
Albany, Oregon U.S.A.
Dearest Isabelle:

My foot slipped and I'm headed for Toronto to join the Royal Flying Corp. I'm sorry, dear, a lot sorrier than I'll probably ever be able to make you believe, but I couldn't resist the old itching foot. The wanderlust got me.

    There are three boys from the south with me and a couple of others are coming up in a few days. We go to Toronto where we get two months of ground school. When we pass our exams there we become second lieutenants and after we learn to fly we are advanced to first lieuts.

I've two or three little things I'd like to have you keep for me if I ever get headed for France. I'll write you about them later.

You've no idea what the war really is, Isabelle dear, until you hit a town like this. Young men are few, uniforms are everywhere, and convalescing cripples aren't altogether missing. The daily papers bring a pretty good sized casualty list and several pictures of home town boys who won't ever come back. The whole atmosphere is different. I can't exactly explain it.

I'm crazy to get to Toronto and to start training. It's going to be a regular adventure. It took me a long time to decide, but I can't help but feel that I'm right. I'd sort of figured that I'd go back to school and become real educated and such, but I couldn't quite do it. This question of going is absolutely individual—every fellow has to figure it out for himself—and here I am.

You'll write to me, won't you, Izzy, and think of me occasionally for this war isn't going to last forever and when it's over I'll be back with my head in the air. I may be off on the wrong foot, but I can't believe it.

Love from,
Bo

. . . .

Toronto
Friday
September 14, 1917

Dearest Isabelle

It's now 7:30 A.M., an unearthly hour, and I've already had an hour of drill, folded up my blankets and put them out, cleaned up my tent and rolled up the walls, straightened up my kit, washed, shaved, brushed the hair, shined the shoes, etc. etc. ad infinitum. Before the day is over I will have had an hour of Swedish drill, two more hours of drill, an hour of wireless, an hour with a Lewis gun, and an hour of map reading. The remainder of the time is mine to do as I please with. Oh yes. Today we also get our pay, a shot in the arm for typhoid, and a swim in Lake Ontario which is the coldest lake known to mankind. The above mentioned spare time is devoted to study, learning to get eight words per on a ticker, learning a positively gigantic list of wireless code signals, getting hep to the way in which aeroplanes cooperate with artillery, memorizing the hundred odd parts of the Lewis gun,—but why carry on? It's a hard, hard life.

Your letter, dear, was the first and, until yesterday, the only one I received in this strange land. I certainly was glad to get it. I'll probably go nutty from lonesomeness before I get out of this place. All the people are so funny and so terribly slow in the head. There are a lot of Americans out at this camp, so it isn't quite so bad. . . .

There are no machines here but there are several flying camps close by and there are planes around all day long. Yesterday was very fine flying weather and there were always three or four in sight.

They are going to move the flying camps to Texas for the winter, and if everything breaks pretty I may get out to those dear [in] California before going overseas. . . .

Write to me often, little lady, and if I don't answer as promptly as I should just figure that it's because "I'm in the army."

Love to you dear,
Bo

. . . .

Long Branch,
Ontario
Tuesday
September 18, 1917

Miss Isabelle G. Young
Kappa Alpha Theta House
Stanford University
California, U.S.A.
Isabelle Dear

This is going to be a hell of a letter right from the start. Last night Pop Taylor brought your last—or latest—letter out from town, and I read it by the light of my little oil lamp. I wasn't surprised—nothing surprises me any more—but I gently cussed myself to sleep and woke up this morning still cussing.

Now I'm going to tell you a lot of things—or at least two or three. The first one is that my name is Bo—get that. Bo. Nothing else. Bogart is a most undesirable name, it's entirely too formal, it takes too long to write, and it's all wrong that you should start being formal at this late date.

The second little thing is that I don't know of any one thing I could use more than a knitted sleeveless sweater, and I'm sure I don't know of anyone I'd rather have make it than your own sweet self. If you . . . fabricate me a sweater I'll wear it in the morning for parade, to lunch, to tea, and I'll even sleep in it. Really tho, Izzy, you have all the instincts and earmarks of a most regular person. And before I knew you I always had the idea that you were terribly cold and distant and such.

Something else I'd like to have you know. I'd like to be in Los Angeles now . . . , and I'd like to be going back to Stanford, but in a good many ways I'm glad that I'm right here. There's something going on—as the young lieutenants around here say, "the big show"—and anyone that doesn't get tangled up in it is wrong. . . . The good Lord knows I wasn't in any danger of getting shoved into the infantry, but things just looked different. If nothing happens I don't think it will ever be regretted.

One more little detail and we dash away for breakfast. If anyone asks you what Rogers is doing you tell them he's flying, get that, *flying*. But just between you and me before he gets off the ground there's some dozen or more arts, crafts, and sciences that have to be pretty well tinkered with. They include wireless, map reading and making, gunnery, aerial gunnery, meteorology, aeroplane construction, motors, artillery cooperation with aeroplanes, night and cross country flying, military law, etc. etc. So before there is any soaring around in the blue there will be a month of regular study and classes. Just like college.

Well, little lady, this about concludes the Tuesday morning "straffing." This note is probably all wrong but so is everything else, so I can't see any particular harm.

Love from,
Bo

. . . .

Toronto
Sunday
September 30, 1917

Isabelle Dear,

. . . Yesterday's orders contained the dates of leaving for Texas, and unless something goes wrong your humble servant will spend the cold months in those dear U.S.A. And if I get that close to home I'm going to spend a few days home and a few in Palo Alto. However I'm a long way ahead of my schedule, and a million things may happen between now and then. . . .

Last Monday an even hundred of us were dragged out of bed at 4 A.M. and packed in here to the University. At present we comprise flight 3 of No. 4 School of Military Aeronautics. Believe me, kiddo, since that eventful Monday morning I haven't had time to sit down and quietly smoke a cigaret. The war is no joke to these birds and speed is the paramount idea

Isabelle Gibson Young, age twenty.

in their heads. They jam enormous quantities of data at you, and believe me they know just where they are heading for, and they aren't losing any time in getting there. They figure on stuffing you just as full of knowledge of machines and flying as they can in six weeks and then on teaching you to fly. After you get your wings and commission on this side you go to England and really learn how to handle a machine. According to the officers here, after you have done your sixty hours solo and all your landings and turnings, you merely know the rudiments and should be ashamed to call yourself an aviator.

The work has been terribly interesting. The instructors are all young officers who have been out in France for two or three years, and their lectures bubble over with personal experiences. Most of them have had a lot, for they are wearing wound stripes on their sleeves, and a couple of them who were pretty well shot up in the legs limp a bit.

I haven't had any particular worries about going up, altho they say that the first time isn't a sensation that people would fight for. I suppose it's very much like anything else when you get used to it. Most of the returned pilots around here have lost their nerve or at least the desire to fly any more. They say that a man gets fed up on flying after a couple of years of the kind of work they get out in France. It's pretty hard on the nerves.

They have a little scout plane that just arrived from England and that is absolutely the latest in fighting machines. It's called a Sopwith "Pup," and its barely twenty feet wide, but all motor and strength. Its maximum speed is about 130 per. You can climb in the seat and imagine all kinds of things—which I very naturally did—but it will be November before I get a chance to break my neck.

Must copy some notes on night flying, so this seems to be about all. From now henceforth my letters are going to be cut down to about one per week. I simply haven't time for more, dear.

Love from
Bo

. . . .

Toronto
Saturday
October 6, 1917

Dearest Isabelle

... Primarily—Monday is thanksgiving in this part of the world—altho God knows there are a good many people up here who haven't so terribly much to be thankful about. We get a half holiday and I guess we're pretty lucky to get that. The C.O. (Commanding Officer) informed us that he was sorry but there was a war going on, and it was a cinch that the usual quota of aviators would get shot up, and it was up to us to help them out.

Then Bill Taylor, (not Palmer) the chap I came up with, applied for a seven day leave to go down to Uniontown (Pittsburgh) and drive in the races there. He holds the speedway trophy which he won in May, and he had a clear guarantee of $1000 and a chance to win a good sized purse, and had a Dusenburg car all ready for him to drive, but the C.O. said that it couldn't be done as they needed men too badly overseas, and if he made an exception in Bill's case he'd have to in others.

And then one of our instructors—a young captain named McMullen—told us all about the latest reports from headquarters about the new "Dolphin" scout, a machine that is making 170 per and climbing to 20,000 ft. in fifteen minutes. And Major [Billy] Bishop, a Canadian who is the high man of the RFC having over forty Huns to his credit and about a ton of medals, says that most of the fighting is going on at 20,000 ft., and that there is even then danger of being attacked from above.

And it has been regular blue weather with drizzly rain and cold winds. I thot [thought] I must be getting cold feet or losing my nerve or something. . . .

I had a hundred things I wanted to tell you—about the letters I've had from the bros. at L.S.J.U [Leland Stanford Junior University]—about the bagpipes that we have to march by now—bagpipes & drums that you could march thru Hell behind—about the sweater and how I appreciate it and how I'm going to be able to use it—about how they have cut down on our ground school course, and how we will probably be flying inside of three or four weeks—about how I suddenly woke up to the fact that Palmer Taylor & Ellen Calhoun had been traveling around together for a long time—and about a hundred other things that might have really interested you.

But I got sidetracked. I'll try to keep on the straight & narrow the next time.

Good night dear,
Bo

. . . .

Toronto
October 10, 1917

My Dear

. . . Haven't had a letter from you in three or four days, and just between you and me that's an awfully long time to go without a letter from you—at least it seems long. I've also got my eye open for the sweater which is going into service just as soon as it arrives.

* * *

Now since that whistle blew about seven hours have slipped by, and I've had an hour of wireless and two hours over a Vickers Machine Gun and have had tea and been for a walk with Pop Taylor, a long session of useless conversation with one of my roommates who, by the way, is named Tabor and who was All American half when he was at Cornell a couple of years ago. I've finally managed to light long enough to finish this letter. . . .

Oh yes! You were to know about the bagpipes. Just where the Commanding Officer picked them up is more than I can say but they're here. They are on the job at early morning parade, and they don't leave until we finish in the afternoon. The drummer waves those two sticks in the most marvelous fashion and the darn things play "The Camerons are Coming etc." and a lot of other fancy music, and you just can't help marching. My last request is going to be bagpipes at the funeral.

We were informed yesterday that we would probably get our exams sometime next week which means we will—if we are fortunate enough to pass—be posted to flying camps in a couple of weeks. They are going to fill all the squadrons up before they send them to Texas, which will probably be very shortly. Things are looking up. . . .

Good night dear
Bo

. . . .

                                                Toronto
                                                Saturday
                                                October 13, 1917

Kappa Alpha Theta House

This is going to be short and not particularly sweet as my noble side kick —Bill Taylor—and I are about to dash out and see a regular show "Have a Heart." We have seats well to the front and expect to have a large evening.

Bill, by the way, is going to do the army dirt, at least he is very likely to, and is waiting downstairs now for a phone call from Pittsburgh to find out all about it. As I may have told you he is the proud possessor of the Uniontown trophy and tried to get leave so he could go down there and drive next Saturday. Since they refused to give him leave, the Uniontown Speedway people have raised the ante to $1000 bucks for appearance money alone and a chance to win a lot more. He is phoning to find out if he can get a car to drive, and if he can he's going to just naturally bust away. The most they can give him when he gets back is a few days confined to barracks and a little fatigue duty so he figures it's worth taking a chance on. 1000 fish is much money. . . .

Now listen, dear. If you had half as much trouble making that sweater as I did sewing a six inch patch on a wing this morning you certainly have done something that I ought to appreciate. For two solid hours I toiled with that little patch and a needle and thread. Under and over and thru and under and over and thru. About every third time I'd jab my finger and the patch looked like murder had been committed on it before I was thru. And then it had to be covered with dope—and dope is the technical name and not slang—and the dope contained just about enough ether to make me groggy. I had a fine morning. . . .

Good night, dear

Bo

. . . .

Toronto
Sunday
October 14, 1917

My Dear

So men in khaki give you a thrill, do they. That's all very well when you see a lot of darn good looking Americans parading around the Palace or the St. Francis—they do the same thing up here, and the King Edward is packed with 'em every Saturday and Sunday, and there aren't only Canadians but Americans from both the army and navy, Frenchmen in their light blue and their red-crowned hats, gaudy Highlanders in their kilts and plaid stockings, and very often a Belgian or Italian officer. Very pretty.

Just after I had addressed your letter and was about to bust away with Bill a good looking young officer who had been writing at the desk behind me turned around and asked me if I'd mind addressing an envelope for him. Then I noticed that his left sleeve was empty, and he couldn't steady an envelope. I did it for him, and the letter was to some girl in New Haven. And the men who limp and the men with one leg gone and the men whose nerves are all busted up. Believe me, Izzy, it takes all the glory and the glamour out of the snappy unies. There are far too many young fellows up here who are all busted up. And you should see the re-education school they are fixing up for the fellows who are going to learn to do something with one hand or some work that doesn't require a couple of good legs. You folks in the states will get hep to what war really is after you've seen a few of these birds.

Palmer [Pop] Taylor came over from his official residence which isn't anything to brag about. He comes over on Sundays and stays all day. When these would-be aviators get together and start telling each other about their respective opinions of the army—the RFC cadet wing in particular—the air simply gets blue, and we have to open the windows and let in a little fresh air. However they all have to admit that they wouldn't be anywhere else. The more I hear from home about the dozens of fellows who have been drafted and who have joined the Naval Reserve, the more thankful I am that I'm up here. Possibly I told you that Winnie has been trying to get transferred to the Naval Air Service. They have a couple of squadrons up here at Borden, and they certainly are a fine lot of fellows. Most of them are from eastern colleges, and I'll be darned if I didn't meet a couple of Kappa Sigma brothers one night at the Khaki Club.

Oh yes. The Khaki Club is run by a bunch of war widows, and you can break in there almost any time of day or night and see someone you know and get a cup of tea and some hot toast and kid a couple of the widows along without losing your reputation. And then you always feel as if you were doing someone a little good by trying to make them forget their troubles. You ought to see me drink tea, Isabelle. I can balance a cup of scalding Lipton's on either knee, hands off, and never spill a drop. And I'm getting into the disgraceful habit of carrying my handkerchief in my left sleeve, and wrist watch. Good Lord, all the things that I used to laugh at I'm doing. Ain't war hell?

These hounds are shouting about pistons, and lubricating systems, and monosoupapes [single-valve rotary engine], and crank pins, and connecting rods, and obdurator rings, and I can't keep my mind on this letter so I'm going to quit. I'll get up some time in the middle of the night and write you a real letter. Write, dear, for—well, you simply have to.

Love from
Bo

. . . .

Toronto
Saturday
October 20, 1917

Dear

The sweater and your very sweet little letter both came a couple of days ago, and I promise you I haven't had time to sit down and tell you just how much I appreciated both. The sweater fits wonderfully, Isabelle, altho when I first took it out of the box I nearly went thru the floor. I thot I'd have to send it home to my small brother, but when I put it on there wasn't a flaw. Just like I'd been poured into it.

. . . [B]less your heart I haven't even been off the ground, and while it looks very much as if we were eventually headed for "that bloody land across the pond" as Smith would say, we won't even finish our training here before the first of the year.

We started our exams this morning, and I'm O.K. so far. It seems just like the end of the semester. We get the worst part Monday and Tuesday, and if everything breaks right we should be posted Thursday. However there is always the possibility of failing to make the weight if I may revert to the vernacular of the street. We are trying to pull a few strings so we can go to Deseronto which is about a hundred miles from here and which is the best of the camps. They have better machines and the barracks are very good. Also we eat at officers mess which is a consideration.

Texas is in the future but nobody seems to know just where. Certainly it won't be far away if we get any amount of snow. Two squadrons have gone down already. . . .

The sweater keeps me very warm, and the letters, too, but differently.
Good night dear
Bo

. . . .

Toronto
Tuesday
October 23, 1917

Dearest Isabelle

. . . [B]elieve me, dear, the good old farm [Stanford University] never produced anything like they handed us yesterday. From 9 A.M. until 5 we took five separate and distinct exams on just about everything we ever had and a lot more. And this morning we struggled with two types of machine guns, took 'em apart and put 'em back together again. It's all over but the shouting, and all we have to do is wait in suspense for a couple of days until the postings are made. Just to show everybody how optimistic I am, I went down and bought myself a trunk this afternoon. Just as if I was going someplace. These are going to be two bad days.

* * *

. . . This morning we went on a beautiful forced march thru the mud for a matter of eight miles. This afternoon we were lined up pretty. I passed, and with thirty other bugs am posted to Camp Mohawk, Deseronto on Friday morning—trunks and kits to be ready in front of the residence at 8 sharp.

Izzy dear, everything seems different now. By Saturday I'll probably be holding on with both hands and wishing that I'd never left home. We were lucky for the rest of the bunch—about two squadrons more—won't be posted until next week and possibly not then. We are stepping right into a vacancy and ought to make fairly good progress. A lot of fellows applied for leave and went home, but all the leave I could have gotten would have taken me about to the middle of Nevada or some such place and I'd have had to turn around and come back. . . .

Love from
Bo

. . . .

Camp Mohawk
Sunday
October 28, 1917

Isabelle Dear

"Eat drink & be merry" for yesterday they crashed four machines and today they will probably crash a few more. However none of them are very bad and the worst thing that happens is for somebody to get a black eye or maybe a busted nose.

We haven't been here for two days yet, and I've been up three times, made a lot of landings and take-offs and seen more flying than I ever saw in all my life before.

There are four squadrons here at Mohawk and altogether about eighty machines. Yesterday afternoon when there was very little wind there were always fifteen or twenty machines in the air and a lot on the ground taxiing around to get the wind.

It's terribly simple, Izzy. You climb in, go like the devil for a hundred yards, and then things begin to drop below you. After you get going you haven't any sensations of speed or any funny feelings at all. It's all so perfectly secure and solid that there's nothing to it. The only thing different from motoring is a terrific wind in your face all the time.

This place is as unlike any place we've been in that there is no comparison. We have very comfortable barracks about a mile from the hangars, and we eat at officer's mess which is considerably fancy with a bill of fare (pardon me, *menu*) and waiters all foxed up in white jackets and regular sugar instead of the quartz we used to get.

I'm on early morning flying and have to climb out at 5:30 and be at the hangar at 6:15. We fly until around 8 and then return and work on motors or fly some more until 11:30 when we have an hour of drill. From 2 until 4 we have classes in gunnery & wireless then more flying until dark. We have to dress up a bit for supper, and after that we sit in front of one of the two big fireplaces and just be sociable, or play bridge, or read, or—as I've got it doped out—go to bed very early. . . .

Your sweater certainly is putting in long hours and, it's just the thing for flying as I can't get along with a heavy coat on. A lot of the fellows wear greatcoats, but I feel like a mummy in one. As soon as the exchequer permits, me for a good leather suit. They are about the only thing that will keep you real warm.

Más en poquito tiempo,
Love
Bo

. . . .

Deseronto
Friday
November 2, 1917

Dearest Isabelle

. . . Things have been happening so fast in the last week that it's been hard to settle down long enough to write a readable letter. Yesterday afternoon I finished my instruction—had about four hours and a half—and last evening the lieut. said "Here it is. Fly it." And I'll be darned if I didn't. Guess I was a sort of pale green on that first flight alone, but today I was up for over three hours altogether and made eighteen landings. All the rolling stock is still intact. Pardon me while I touch wood. This morning it was wonderful as the ground was covered with snow, and the air was like velvet. The old machine insisted in climbing so we went up about 3000 feet and wandered around a bit. This afternoon it was awfully bumpy, and I had a pretty strenuous time trying to keep level. It's hard work and not much fun. I am getting so that I can tell how the bumps are going to be by looking at the ground. You see, dear, plowed ground thaws and sends up warm air. You hit an ascending current and get bumped up. A cool grove of trees or water act the reverse.

Tuesday was a bad day, windy and rough, and there were two bad crashes. A couple of fellows came together at about 150 feet, both looped backwards and splintered a couple of perfectly good machines. One of the boys crawled out without a scratch, and the other was just cut up a little. About an hour later another beginner went into a pretty little side slip right in front of our hangar and busted up another buggy completely. He sort of wiggled out and wasn't badly hurt aside from a few cuts. However about $20,000 worth of neat woodwork and motors went to the junk heap. This afternoon the place looked like a battlefield, machines on their noses, machines on their backs, and nobody hurt.

One word for the sweater. It works as a sweater from 5:30 A.M. until

5:30 P.M. I don't wear it to supper but from 9 P.M. until 5:30 A.M. again it serves as a covering for my pillow which, I am sorry to say, isn't too clean. We don't seem to have pillow covers. . . .

Love to you, dear.

Bo

. . . .

Toronto
Tuesday
November 6, 1917
King Edward Hotel

Izzy

I feel like a lost soul this evening and plenty of reason. In the first place it's past midnight and I just arrived from Deseronto and I'm all alone and tomorrow at 9:15 I grab a train for Camp Borden and enter the 42^d Wing, RFC, sans future, sans friends, sans everything. All the fellows I went out to Mohawk with are still there. I passed out of the Elementary Training Wing this morning having completed the necessary ten hours solo flying and the forty landings. I certainly hated to go but glad to go to Borden as they move south on the 15th. Looks as if I finally get to Texas. This is a great system, to bust into town about midnight and out again the next morning. However, I would join the army.

Good night,

Bo

. . . .

November 10, 1917
Camp Borden

Dearest Isabelle

Haven't time for more than a page or so as I'm about to dash down and take my altitude test.

This is a terrible joint, cold and deserted and they haven't enough machines to go around and the ones they have are all wrecks, and I don't know anyone here, but next Wednesday or Thursday we climb aboard the old choo choo and head for God's country.

Now here goes a strange and unusual request—don't write any more until I send you our Texas address. I have enough trouble as it is getting your letters, and if we get down there and you address them up here Heaven knows what will happen. When you do get the correct address—

WHEN YOU DO—Anyway, you can rest easy for a few days and concentrate on Red Cross, your *other* war duty.

Pop Taylor, who is out at Mohawk, sent me a bunch of Dippys [*Daily Palo Alto*, the Stanford University newspaper] and I learned a lot of campus scandal. Saw your sweet name spelled incorrectly.

About the only person I know very well here, a fellow named Robbins from San Diego, went out last night on an altitude test and hasn't been heard from since. He had a crash yesterday morning and may have "had his wind up." He swore he was going to do some stunts and a couple of loops if he could get up enough nerve. I'm sort of worried about him.

Here I go.

Love to you Izzy dear

Bo

. . . .

Camp Borden
Tuesday
November 13, 1917

Izzy Dear

... Thank the Lord we leave here Thursday morning, and with any kind of luck should be in Fort Worth by Sunday. We are going in three trains and thru Detroit, Chicago etc. It certainly will seem good to get among human beings again.

Took my instructor's test yesterday and nearly got lost in the haze and dark. From 3000 ft. you could barely see the ground, and I had a terrible time locating the aerodrome. Busted a tire when I finally did land. If there is a decent machine around this morning I may get a chance to go cross country to Toronto which is about eighty miles. Certainly would like to go.

The chap I told you about who hadn't returned Saturday got lost and crashed a machine in some farmer's barnyard. Didn't get hurt a bit, however. They have been breaking up buggies so fast during the last week that they hardly have enough to carry them over today which is the last day of flying.

I suppose I told you that Palmer was at Deseronto and will probably go south at the same time we do. He may be in the same wing there as some of the Deseronto squadrons are going to be with us. I certainly hope he is. He gets Dippys.

It's almost ten o'clock, and I have to dash away to the range and shoot a few clay pigeons. They say shooting the darn things improves the eye for Huns and you can bang away at them to your heart's content. However, I imagine it's much tamer than shooting at Germans. The pigeons can't shoot back.

    Love,
Bo

2. TEXAS *November 1917–January 1918*

As Bogart Rogers completed his elementary training in the frigid Canadian winter, the Royal Flying Corps agreed to train United States Air Service pilots in the winter sunshine of Texas. Corps cadets traveled to Fort Worth for further training with Air Service student pilots.

The Texas winter proved unpleasant, and the unfinished state of the camps made training difficult. Measles, spinal meningitis, and scarlet fever struck the training camps. General William C. Gorgas, chief of the United States Army Medical Corps, who was noted for having eradicated malaria and yellow fever during the construction of the Panama Canal, was brought in to stem the epidemics.

. . . .

Fort Worth
Monday
November 19, 1917

Isabelle Dear

So this is Texas!! Well, well. Anyway, it's God's country again and that helps a little.

We got here yesterday after a somewhat tiresome three day trip. We had good accommodations and lots of good grub but we whistled thru Detroit, Chicago, and St. Louis and didn't even get a chance to poke our heads out of the windows. Our only two stops were at a little one-horse town in Illinois and at Fort Smith, Ark. In both places we made route marches, had about an hour in which to buy a lot of junk, and we were always royally treated by the native population. They presented us with assorted food, candy, cigarettes, tobacco, etc. etc. The sight of so many foreigners seemed to awe them.

I had a section with an American cadet named [Harry] Jackson, a Yale '17 man and we were across from two other Americans, one a Yale '19 boy

*Cadet Bogart Rogers in front of a Curtiss JN-4 Jenny,
Camp Benbrook, Texas, November 1917.*

named Corse, and the other a Michigan fellow, Smith by cognomen. We played poker, fantan, solo, rum, hearts, bridge, every card game known to man, had a high powered quartette going, just sat and talked and did everything possible to pass away the time. We passed considerable. They were three fine boys and we have been together here.

This has the makings of a wonderful camp, but it's a long way from completed. We have been sleeping in half-finished barracks, haven't any lights, heat, running water, and eat periodically of field rations—strong & greasy bacon, beans etc. etc. The camp is wonderfully arranged, and the aerodrome is over a mile square. There are twelve hangars, four cadet barracks and four for the mechanics, six mess halls, a lot of repair shops, officers quarters, in fact everything will be fine when it is finished. Our barracks are only about thirty yards from the hangars, all of which is very fine for early morning flying as you can crawl out at 6:28 and be on the job at 6:30.

We probably won't get any flying until the first of the week. Capt. Castle (Vernon) [who had a famous dancing partnership with his wife, Irene] was up for a while this afternoon doing a lot of stunts—loops, spins, Immelmann turns, and all the old stuff. It is a perfect place for flying as the weather will probably be fine most of the time, and there isn't a tree within miles. The machines are all new and have the very latest improvements and should be great to fly in. I'm afraid I'll have to get reckless some day and try a few stunts.

The last letter I wrote from Borden had a lot of flying before I mailed it. Right after I wrote it I went down to the hangars, and was sent cross-country to Toronto. Went down—about seventy miles—in an hour, landed at two aerodromes, got my papers signed, had lunch, and returned to Borden. The whole trip only took a little over three hours, and it was a wonderful ride. I could see Borden from fifty miles away and that at about 6000 ft. The air was as smooth as glass, and there wasn't a cloud anywhere. Later in the afternoon I went up for my altitude test and climbed to 8100. Certainly is fine up there, and the higher you go the better it is. You'd be crazy about it, Izzy, and some day we'll go for a joy ride in somebody's machine. I'm afraid it won't be mine at $8,000 per. . . .

Good night little lady
Bo

. . . .

Fort Worth
November 26, 1917
Camp Benbrook

Isabelle Dear,

Right here and now I want to thank you for the candy. It was probably the finest box of candy that ever arrived in this part of the country. The reason that I know it was a fine box of candy is because the tramps that ate it told me so. Permit me to explain.

I've probably told you before that the RFC has the bummest mail system know to man. Furthermore in the last couple of weeks I've been in about every squadron in the place. First it was 87 Squadron in the 43d Wing. Then I got posted to the 79th Squadron in the 42nd wing, then to 78 Squadron, and when we came down here I was posted back to 83 Squadron in the 43d Wing for higher training and then to 81 Squadron. Furthermore from the time I write to you until the answers come a couple of weeks pass. Very well. You follow me? Fine! There are three camps here at Fort Worth. I am at Benbrook, Pop Taylor, Bill Taylor, all the fellows I know or did know are at Everman. For over two weeks I've not had one letter. Today one of the boys flew over to Everman and I asked him to see if he could find any mail for me. He did. This crook Pop Taylor had it—a lot of old letters some dated back almost a month. These crooks had the nerve to get my mail, saw the candy, and darned if they weren't kind enough to put your card in an envelope and send it to me. I'm enclosing another letter on which the burglars jotted their individual opinions of the candy.

[enclosure]
THE OTHER LETTER
"Just received a box of candy from Isabelle. Many thanks, Bo." "Same to you, Bo. Billy." "It sure was fine. Pop." "Came just in time. Tom." "Swell food. Marry the girl. Pat."

Izzy, dear, at first I was mad as the devil and smoked up the pure atmosphere with some swell cussing. Then I laughed probably because it was the only thing to do under the circumstances. We are all going to have Thanksgiving dinner together and how those birds will give me the ha-ha.

Don't know just how long we will be here but probably until the first of the year. Things are still going slowly and we aren't doing an awful lot of flying. I was up Friday on my grand signal test and certainly was busy trying to see the signals, fly the machine, and send down wireless. The air is awfully "bumpy" and keeps you pretty busy. Tomorrow I'm going to do my photography if it is clear enough. We have to fly all over the country and take pictures of about twelve points, roads, buildings, etc.

Good night, dear
Bo

. . . .

>Camp Benbrook
>Thursday
>November 29, 1917

Dearest

Thanksgiving here passed about like any other day only we had grape fruit for breakfast and turkey for dinner. There was the customary flying and the customary classes. This evening we had a very interesting talk by a Lieutenant Hearn who has just come from France.

Unfortunately the whole camp is confined to barracks, as there has been a pretty serious outbreak of measles and spinal meningitis around Fort Worth. Camp Bowie is quarantined, and all the other camps are confined so we will probably have to exist on the army grub and enjoy the scenery in the immediate for some time. . . .

* * *

A chap I knew at ground school was killed at Everman a couple of days ago. Maybe I told you about Bill Taylor. The fellow I came up here with? Well, Bill, an American Cadet named Porter, and a boy named Alcock were flying formation.

>Porter
>+
>
>Taylor   Alcock
>+   +

Alcock had the fastest machine and kept flying "S" turns to keep in place. On one of his turns he slipped under Porter and tore his (Porter's) undercarriage off, broke his right lower wing, and jammed his ailerons. Alcock's machine simply collapsed and fell a neat 3500 ft., and his gas tank exploded. Pretty mess. Porter, a skinny little kid, never lost his bean for a minute, and with only his elevators and rudder working got down and pancaked nicely in a field. Poor Bill Taylor was only about twenty-five yards away when the whole thing happened. I guess his wind was up a little afterwards. Anyway they sent him east with Alcock's body.

Another chap got it at Camp Hicks the same day doing a loop. He came out of it in a straight dive and pulled out of it too quickly. Result: snapped off both his wings. He was about 4000 [feet] up. I was going to try a loop some fine day myself but I'd sort of changed my mind. There were a lot of minor crashes yesterday, as the field was muddy and it was very easy to go over on the nose while landing. It really is funny to see them flop over, and the poor unfortunate drop out just as far as his belt will let him and hang there, or when they go on the nose to see them climb out and look the machine over.

Speaking of the mud—Texas dust is the dustiest dust in the world. When the wind blows you carry a large part of the state around in your hair, ears, etc. But when it rains you carry it around on your feet. It makes a regular nest around the king's boots, and you simply can't get rid of it. It is the most tenacious, the most persistent, the most affectionate, in fact the stickiest mud known to man. Enuf of mud. . . .

We moved into new quarters yesterday—new and very fancy. They are in one wing of the officers quarters and have suites of rooms, four to the suite. I'm with three other fellows, one of them incidentally, is named Fuller and had a brother who went to Stanford. Possibly you knew of him. He was pledged Phi Delt in the '19 class and had only been there for a couple of weeks when he died. I'm still with one of the Yale boys, but the others were sent to Everman. The fourth member is a Beta from Michigan.

I don't know just what I'm going to do for the next two weeks, as I'm thru with my higher training and the next gunnery school doesn't start until the 15th of December. I'd like nothing better than to stay right here and fly, but when you are thru with your work it's impossible to get hold of a machine. They don't encourage joy riding.

The higher training work was very interesting. We had to do photography, bombing, panneau, puffs, and ground strips. Permit me to explain.

Photography consists of taking a series of pictures of landmarks around here—trenches at Camp Bowie, silos, crossroads, and a lot of others. You have a special camera bolted to the outside of the fuselage, and you have to look over the side of the machine to sight the pictures. While you are trying to get a line on your object the machine does all kinds of funny stunts. Ground strips consist of reading signals placed on the ground and sending down answers by wireless. Panneau is reading Morse code from the ground. You use a Klaxon horn to signal down and have a great time. The day I went up there were a lot of machines on the road outside the camp, and when I'd fly down and toot the horn they'd all toot back.

Bombing is just that only, you don't drop real bombs and you drop them by means of a bomb sight, which makes automatic allowances for wind, speed etc. It's comparatively simple to hit your target with them.

Puffs is the imaginary ranging of guns on targets. You know what your target is—usually a crossroad or building, and when you send wireless signals down smoke bombs are fired off at different points, one at a time, and you send down corrections just as if it were regular shell fire. The method used is the clock code—get some L.S.J.U. soldier to explain it to you. It's something like this.

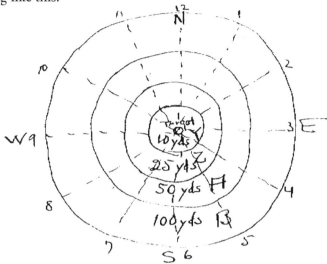

The target is the center of an imaginary clock—12 always being due north. Therefore a shot due south would be at 6 o'clock. The distance from the target is shown by circles Y, Z, A, B, C, etc. Therefore you can send any correction by simply giving the time and the letter of the circle within which the shot falls Simple?? Clear as mud.

Love to you, dear
Bo

. . . .

Fort Worth
Wednesday
December 5, 1917

Hello Dear

Just a few hit and miss remarks before supper which is due very shortly. I'm all thru, dear, no more solo flying for the present and nothing to do but wait for gunnery to open next Monday. I've qualified for my wings and after about three weeks of gunnery will be ready for the old stars.

There are three of us who finished today, and all we've been doing is to keep away from parades and extra fatigues which is some job. . . .

Saturday or Sunday we will go over to Camp Hicks and get settled. Monday the work begins. . . .

My gang's headed for the mess so here I go with them.

Love to you from

Bo

. . . .

Camp Benbrook
Saturday
December 8, 1917

My Dear

. . . Yesterday about noon one of the justly famous Texas northers set in very suddenly. The storm signals were put out, and two machines crashed trying to get down. The wind increased, and we spent all afternoon playing poker and trying to keep warm. By sundown it was uncomfortably cold. We got about fifty candles and put them all over our suite— get that—and continued playing poker until bedtime. Incidentally your sweater was instrumental in keeping me from freezing during the night. This morning it was 12°, and all the pipes in the place had frozen tight. There was ice every place that there were any little pools. Icicles hung from all the faucets, and the wind howled. Being an exceedingly sanitary creature I must wash the face and brush the hair before breakfast. There was a faucet over in front of one of the hangars that was still working. Then I began to notice how cold it really was. I washed, wet my hair, and dashed for the barracks. When I got there, my hair was frozen.

They put all the carriages to bed last night without letting the water out of the radiators, and there wasn't one of them that would run this morning. Believe me, this cold weather is the bunk. Canada was a summer resort compared to this place.

We go to Hicks either tomorrow or Monday and start gunnery. Nobody seems to know just how long it will take, but I figure on being thru— not home—by Christmas or very shortly thereafter. . . . Gunnery promises to be very interesting and not too easy. We do a lot of work on the range, firing from the air at stationary and towed targets, battle practice, camera gun practice, and plenty of lectures on fighting, "stunting" and everything we are supposed to know overseas. We won't get a chance to do much flying and practically no solo work, but I guess they'll keep us busy.

The quarantine continues, and the measles catch somebody every day or so. They are really having a bad time at Bowie. Every day ten or twelve die of measles, pneumonia, or spinal meningitis. The base hospital is full, and the sick list seems to increase steadily. . . .

So your French lieutenant caused a sensation with all his wound stripes etc. There are so many of them around here that nobody ever notices them. One chap, a Captain Cupelle got his commission by crawling out on the wing of his machine at 10,000 feet to balance a broken plane while his pilot brought it down. Another one, Lieutenant Hearn, was in France as late as August. He was flying scouts, but hooked up with a "circus" one morning and came down with fourteen machine gun bullets in his legs. A Capt. Williams got his doing photography considerable distance up. The first thing he knew his arm was full of lead and his observer brought him down after he had fainted. You sort of get hardened to the crosses and the stripes and everything concerned with war.

Tomorrow we get stoves, electric lights, and probably the pipes will thaw out. Everything will be pretty. And when we get nice and comfortable we shall move away. It was ever thus.

Good night, dear,
Bo

. . . .

Camp Hicks
Tuesday
December 11, 1917

Izzy Dear

In as much as it wasn't much colder than a mere 32° all day I'm sufficiently thawed out to write, and naturally you lead the list. When you go to church next Sunday—if you aren't too busy studying for finals—pray there won't be any more Texas northers—at least not until I get out of here.

Sunday we moved away from our nice warm stove at Benbrook and came over here to gunnery. We sleep in an empty hangar on straw mattresses, three blankets, and all the ventilation in this part of the state. Some fine morning I'll wake up frozen stiff and cold and thank my lucky stars that the place I'll go to will at least be warm.

However the mess is a decided improvement, and I'm back with the gang I came from Toronto with. I've been spending most of the time with a chap named Harry Jackson who I probably told you all about before. He certainly is a wonder, and I hate to think that we'll soon get pried apart. But that's the way in the army.

It looks very much like we would be quarantined here over Christmas if we don't get thru before then. General Gorgas, the head of the U.S. Medical Corps and the man who got his reputation at Panama, is here in person to handle the trouble at Camp Bowie and has continued the quarantine there indefinitely. That probably means we will be included, as there is a new case of measles every day or so in the camp.

We have had a lot of interesting lectures on aerial fighting and such and some work with the guns. In a week or so we get a chance to fire them from the air both at towed targets and at ground targets and to have battle practice with camera guns. You simply chase another machine around until you get a good shot at him and then pull the trigger. Instead of bullets you get a picture of him that shows how good your shot would have been.

At present we are in the canteen, about the only warm place around here.

Must quit, dear, as there are lots of notes on ring sights to be copied. You're going to write me a nice Christmas letter, aren't you, Izzy, and I'm looking forward to it.

Good night, dear,

Bo

. . . .

Camp Hicks
Tuesday
December 18, 1917

Dearest

Now Izzy dear you aren't half as worried about these epidemics as I am. This one here is comparatively harmless alongside of the Stanford one. That is really serious. . . .

From all appearances we leave Sunday for Toronto and our commissions and leave—just how much I don't know. I'm afraid to even wonder, for if we don't get enough for me to come to those dear Pacific Coast—Don't you understand by now, dear?

The wild northers have ceased, and in their place has come fog, a warm muggy fog, not like San Francisco. There has been no flying for three days, but we've spent the time listening to lectures and shooting clay pigeons and such. The lectures are wonders, but make you sit up and wonder just how much chance you're going to have when it comes to loops, and spins, and side slips, and tail wagging, and a million other stunts that are essential overseas. I've got to learn a lot more about flying than I do now or some German is going to get an Iron Cross. . . .

This overseas stuff doesn't look quite so alluring as it did when it was further away. However the service machines are wonderful—light[,] powerful, and strong, and very light on the controls. I guess it has its advantages. . . .

A very Merry Christmas to you Izzy, and my love,
Bo

. . . .

                                                 Camp Hicks
                                                 Friday
                                               December 21, 1917

Miss Isabelle G. Young
Albany
Oregon
Dearest

Today has been a wretched day foggy and cold and with the wind back in the north. It was a complete "washout" as far as flying was concerned, and we had the whole afternoon off. We played bridge most of the afternoon. After that I came back to barracks and broke one of my few rules of conduct—ie and to wit—got mixed up in a crap game. Now in the army there is a crap game going almost any place and almost any time. They are exceedingly easy to start and all the material necessary is a pair of dice and some of that well known stuff called "jack." About once a year I make a firm and cast iron resolution never to shoot craps again—and that resolution is always made after I have been effectively and thoroly cleaned. I don't know how in the first place, and the dice won't work for me in the second place, and lastly I don't seem to be able to get very enthusiastic over the game. But the game was going on right at the foot of my bed, and my foot slipped. At present I have just finished making my annual resolution and am sadder, wiser, and far poorer than I was yesterday.

From all I can find out we don't get away Sunday for there has been practically no flying all week, and the aerial work is way behind. Yesterday morning I was up for about an hour doing camera gun practices, but that is the only flying I've done in a couple of weeks.

I get to thinking every once in a while what a pretty future is ahead. Honestly dear, I don't see a prayer for myself. The odds are all wrong and—unless I'm able to learn a lot more about flying than I know—curtains. France—the whole western front—is a hell in the air and even if you are lucky and don't figure too well. Incidentally three boys got done-in in one crash at Benbrook this morning. The left hand circuit flag was up, and one boy forgot and turned to the right. The result was that he ran into an instructor and cadet doing dual.

Guess I'm getting morbid. But can't you see, Isabelle, that there are a thousand things I'm crazy to say to you that I simply can't?

Good night, dear,
Bo

. . . .

Dallas
The Adolphus
Christmas
December 25, 1917

Miss Isabelle G. Young
Albany
Oregon

Greetings—Greetings—Greetings

... Poor old Pop is laid up in the hospital with a wicked cold and very bad eyes. I've been down to see him several times and taken him some stuff, and he wasn't very happy over spending the day in bed.

Yesterday was a bad day. We saw three fellows get done in. Jack [Harry Jackson] was just climbing into a gunnery machine, and I was standing talking to him when a mechanic working on the engine slipped and fell into the propeller. The poor devil never knew what hit him. It wasn't a pretty sight by any means. And then about a half an hour later an instructor and a cadet had a machine explode about 3000 ft. up. That wasn't particularly nice to watch. They fell like a comet. As we knew the cadet pretty well, it sort of got to us. However I guess it's all in the game.

Good bye, dear,
Bo

. . . .

Camp Hicks
Friday
December 28, 1917

Isabelle Dear

I'm enclosing some odds and ends. The newspaper clipping is bunk but may interest you. Fact is I've known Beanie Walker for years and years and known him pretty well. I never wrote to him, so he must have gotten his information from my brother-in-law [Ivan St. Johns, a publicist for the *Los Angeles Examiner*] who runs around with him. I tell yuh! it's great to have a press agent.

[enclosure]
Blinkey Ben at
Camp Kearney
by M. M. Walker

The Ole Night Editor,
Los Angeles "Examiner"—
Dear John:
This aviation thing is to play a wonderful part in the Big Game, John, an' that reminds me, I had a letter from Bogart Rogers. He's with the flyers at Toronto. Out o' a class o' a hundred an' fifty cadets he took second place in his examination, finished with 198 out o' a possible 200. An' it seems like only yesterday that li'l Bogart, in knee pants an' his shirt sewed on fo' the summer, used to sit on the bench with me at Chutes Park watchin' Uncle Frank Dillon's Angels.*

I'm all thru, dear, and have been posted to headquarters. The commission and shoulder strap and little stars are waiting in Toronto, but as usual we haven't the least idea when we are going. Probably some time next week, but you never can tell.

Harry Jackson and the rest of the American bunch have been posted to the 28th U.S. Squadron and will leave for overseas about the middle of January. Seems as if every one was headed in that general direction. The American squadrons will probably go to England for more training and to be equipped with British machines and ultimately will be attached to British wings as complete squadrons.

One of the fellows I roomed with at Benbrook died yesterday of scarlet fever. The poor kid had wretched treatment and should have pulled thru with any sort of decent treatment. These camp hospitals are good places to keep out of.

Pop has the measles. The poor boy is confined in the contagious ward and can't see anybody. I've been over every day and get to talk to him thru the door. It isn't anything serious but pretty hard luck. . . .

Good night,
Bo

*Portion of a clipping from the *Los Angeles Examiner*, November 7, 1917.

. . . .

                                             St Louis
                                             Hotel Statler
                                             Thursday
                                             January 3, 1918

Izzy Dear

I'm all broken up, dear, and I've only a moment in which to write. Palmer [Pop Taylor] died yesterday morning just before we left of scarlet fever. If Ellen [Calhoun] doesn't know for Heaven's sake break it to her easy. I'm going to write to her tonight. It all happened so suddenly that I can't seem to realize that he's gone.

I've been trying to write to his sister for an hour and made an awful mess of it.

I can't write any more, dear. May be home in a few days and may not.
Love to you, Isabelle,
Bo

. . . .

                                             Toronto
                                             Monday
                                           January 21, 1918

Dearest

There are so darn many things rattling around in my head that I'll never tell you all of them in this letter or the next or the next, but they'll all get out eventually.

Izzy, I'm the happiest person in the whole wide world and absolutely no exceptions. Your telegram came a few moments ago, and I was afraid to open it.* If it had been any different I shouldn't have cared what happened to me. As it was I came within an inch of waving the arms and shouting and making myself conspicuous generally. Anyway, I did all that mentally and internally and up my sleeve and in the privacy of my boudoir.

Someday you'll understand dear, how much your love means to me, and I hope how much mine may mean to you. I've thot one thing, and I've thot others, but always it has been the same old conclusion, the same arrival at the same place. I hoped and prayed after my peculiar fashion that I'd be able to see you and ask you what I asked you tonight, but things just broke

---

*Bogart proposed to Isabelle in a long-distance telephone call from Toronto to her sorority house. Since she was too flustered and embarrassed to answer in the presence of all her sorority sisters crowded around the phone in the Theta house, she sent him a telegram in which she accepted his proposal of marriage.

wrong for that. Then, when I had ever come home—well, you may have a hunch—a flash of female intuition—as to what would have happened.

This war bride stuff is wrong, Izzy, all wrong. There's no use in stalling, and the overseas game is no joke. It's all very well to marry some girl and then dash away to France. And if you get back everything is wonderful. But if you happen to be unfortunate, you must see what it means to the girl. It's so much harder to forget then.

Aside from the incidental fact that I love you, there are dozens of reasons why I shouldn't dare ask you to marry me. I'm absolutely ignorant so far as scraping together a respectable living is concerned. I never earned more than six dollars a day in my life, and I'll be hanged if I can figure myself as worth any more to anybody. I've a mother and two small brothers that I may have to do some looking after someday. They're well cared for now, but father—who I might as well tell you—is a very short sighted person and refuses to think seriously of the future. Just incidentally he has a wife of his own to look after. Yes—divorce.

But I don't care. Just as long as I know your love is with me in everything I do, things will come out right somehow.

Gee Isabelle, I'm excited and this letter is about as coherent as the Bolsheviki ideals. My latest desire is to throw great gobs of ink all over the place and hammer perfectly strange men on the back.

Do you want to keep the large secret a secret, or must I buy candy for the hungry sisters. Maybe for a while at least it will be best as it is. You and I know, and personally I can't see why anyone else should know. Of course there are probably a lot of people who might like to know—mothers and fathers for instance—but it ought to keep. It's such a wonderful secret. You tell me just what you think, dearest, and we'll sort of discuss it at long, long range.

Tomorrow I'm going to send you my pin and my ring. The ring isn't much as a woman's ring, but I'm going to write to a jeweler I know in the city, and maybe he can figure out a regular setting for it.

I'm enclosing some of the clever work of my press agent. Aside from the fact that the underlined facts are erroneous it isn't bad. You know I'm not nineteen. Heaven forfend!!! I know it only cost father about fifteen bucks to phone to me, and everybody knows I've been up here over four months.

[enclosure]
Los Angeles Examiner
    EARL ROGERS' SON OFF TO FRONT AS BRITISH AVIATOR
Called back to Ontario by telegraphic orders while he was speeding home

to Los Angeles on a furlough, Bogart Rogers, the *19-year-old* son of Attorney Earl Rogers, member of the British Royal Flying Corps, was intercepted by his father at Chicago and told by long-distance telephone to hurry back to his station to obey orders to embark at once for France.

Several cipher telegrams from the Royal Flying Corps officials were read to the young aviator by his father over the telephone and Attorney Rogers paid a *$35 phone bill* for his talk with his son.

While keenly disappointed that the lad was not able to continue homeward and enjoy a short visit, at least, before going off to war, Mr. Rogers did not attempt to conceal a certain degree of elation that his boy had "made good" so rapidly in the aviation service that he is called to active duty just *three months after* he entered training.

The wings are dirty and they've been worn—by me—but maybe you can clean them and use them for something. I'll be darned if I know what.
Good night, dearest,
Bo

. . . .

Toronto
King Edward Hotel
Tuesday
January 22, 1918

Hello Dear

Just as happy as ever but not quite as excited. I'm actually rushed to death, as there are a million things to be done and very little time in which to do them. I always leave everything to the very last minute and then tear madly about trying to make up for lost time.

Isabelle, the beautiful mess I've made of things during the last couple of weeks has done me more good than anything that could possibly have happened. If I hadn't gummed the deal so completely chances are that I'd never have come-to.

When we arrived here from Fort Worth, we were told that our leave was indefinite but would probably be about ten days. That wasn't very much but I thot I'd at least go to Chicago, so in case the leave was extended I'd be that much nearer home. I stayed with a chap named Gordon Seagrove, a boy I used to know on the coast, who is now Sunday editor of the Tribune. The night I arrived in Chicago a swell young blizzard was in progress, and all the trains were tied up. I must have gone completely insane, dear. I kidded myself into thinking I didn't want to go home, I falsely reasoned that the trains might be delayed for a long time; I said I wouldn't have time; and Gordon and I proceeded to have a grand time—at least it seemed grand then. After about two days of it I woke up. I knew if there was any possible way of doing it I ought to go home. I wired headquarters to find out how long leave would be. They never answered. Then on Thursday night I wired home to see if there were any orders there. Father immediately phoned that there were two telegrams saying to report at Toronto immediately. During the whole time I hadn't written to you or to mother or to father. I sort of thot I'd get home.

Well, I couldn't get a train to Toronto until Monday morning. They weren't running. Tuesday afternoon I busted into headquarters and was told that leave had been extended one week. Izzy, I could have murdered and maimed men, I could have quaffed the old carbolic without a regret. I could have wept crocodile tears and raved and torn my hair. What I really did do was to leave for New York on the first train.

That's very briefly the reason I never came home. But that's a detail. We went on a lot of parties, were with several young ladies, theaters, cafes, regular damn fool time, and thank the Lord that was what woke me up. I didn't have a good time, dear. It's impossible to really enjoy yourself with a lot of people who happen to admire the cut of your uniform or think your wings are pretty or just like to trot around for excitement regardless of who they may be with. When I got to New York and had time to think it over, I could have done a brodie from the Woolworth Building. It was terrible, dear, not to get back to see you and my folks and all my real friends. And I never had intelligence enough to realize it.

Believe me, it's all as plain as day now. And to know that you really care in spite of all the crazy things that have happened is going to make everything so much easier. I can't help but make good now.

Love from
Bo

. . . .

Toronto
King Edward Hotel
Tuesday Night
January 22, 1918

Dear

"The shouting and the
tumult dies.

> The captains and the kings
> depart—"*

All of which translated into plain English means that Rogers has packed his grip, paid his hotel bill, put in a call for six-thirty, and written several letters. . . .

The grand time in New York is a long, long story. I stayed with some old friends of the family, a chap named Harry Mestayer and his wife. They have an apartment on Fifty First Street right off of Broadway and very centrally located. Harry got me cards to the Friars Club and the New York Athletic Club and showed me all over the place. He knows lots and lots of celebrities and showed me most of them. I met K.C.B. George Herriman, the guy who draws Krazy Kat, George Cohan, Jim Corbett, David Belasco, Harrison Fisher, and a lot of others. Then I ran into Walter Morocco who I used to know out west and who is a great party player. What should he do but get me a card to the Lamb's Club and take me to a couple of good shows. We also went on a party at the New Amsterdam Room with Jack Pickford and his wife. Jack is a very nice kid but the wife is a mess. . . .

Aside from that I saw most of the town—it was my first trip there—Grant's Tomb, Clairmont, Central Park, Washington Square, Wall Street, Battery Park, Brooklyn Bridge etc. etc. However, I can't really rave over the place. If you happen to be blessed with a million dollars or a million friends you may have a good time. Otherwise no.

Bill Taylor and I went out to Mineola with Ralph De Palma to see the Liberty Motor. That dago can certainly drive a machine, which after all isn't strange. He's one of the most attractive chaps I ever met. Incidentally, Long Island can't hold a candle to the well known peninsula. There are some fancy looking golf courses and a few beautiful homes but the whole place is crowded just like the City, and Garden City looks like Redwood.

A world of love to you,

Bo

2ᵈ Lieut. Bogart Rogers R.F.C.

---

*Rudyard Kipling, "Recessional."

3. ATLANTIC CONVOY
January–February 1918

On January 23, 1918, Bogart Rogers sent a telegram to Isabelle Young.

> GOOD BYE DEAR. WILL ALWAYS REMEMBER.
> BO

He then departed on the train for Halifax, Nova Scotia, to embark for England. On December 6, 1917, only seven weeks before Bogart arrived, the Norwegian ship *Imo*, sailing out of Halifax harbor and bound for Belgium with relief supplies, had collided with the French munitions ship *Mont Blanc*. The *Mont Blanc* caught fire and blew up. Fragments of steel hailed on the city, killing 1630 people and injuring thousands. Not a building was left undamaged.

Rogers was to embark on the transport *Tunisian* for Liverpool. The *Tunisian* would sail in a convoy for protection against German submarines. In all the convoys bound for Europe, only two transports, the *Tuscania* and the *Moldana*, were sunk, with a loss of 222 lives. The *Tunisian* and the *Tuscania* departed for England in the same convoy that January.

. . . .

> Montreal,
> Queen Hotel
> Thursday
> January 24, 1918

Dear Child

We steamed into this ancient town last night and are leaving for Halifax in a very few moments. I crawled out of my berth too early and had a long walk before breakfast. It really is a very beautiful city all the churches and buildings being characteristically French. Almost all the population is French also, and you hear the natives jabbering to themselves on the street corners and talking with their hands. Aside from the fact that I looked at the thermometer and found it was twenty below I had a fine walk. . . .

Izzy, I know just how deeply you must feel for Ellen. I know about how Pop felt toward Ellen and her folks, and I know that he used to get letters from her and her mother almost every day. Somehow it doesn't seem right that he should have to go. From what Billy Taylor told me, Pop's guardian came south to see about everything and I understood he was to be buried somewhere in Ohio.

Love from,
Bo

. . . .

Halifax
January 25, 1918

'Lo Sweetie

"This is the forest primeval."

All of which may be a trite and perfectly silly way of starting any letter but which is the truth just the same. All day long we've been rolling thru Nova Scotia, and in an hour or so we reach Halifax. This is a wonderful country dear, and it just bubbles over with history and romance and all kinds of interesting little things. . . .

Nova Scotia is certainly the home of Christmas trees and snow. All the country is hilly and rolling and there are miles and miles of woods and the snow and ice is everywhere. . . .

At the last station we passed a couple of trains full of Chinese coolies who are going over on the same boat with us. There must have been 2000 of them. They use them as laborers on the front, and they are said to make splendid men.

Also at Moncton we passed a large German prison camp all fenced around with barb wire and with sentries walking guard. There were a great many prisoners walking about—most of them being interned civilians, but also a great many sailors and officers.

Haven't the slightest idea when we sail. It may be tonight or it may not be for two or three days. Nobody on the train knows anything about it.

There are a lot of Equipment officers with us and they aren't very popular. As a general rule they have "safety first" jobs and flying officers have a sort of contempt for them. There isn't really any reason for it, for there aren't really any bomb proof jobs except those at home. The fellows who work with the Mechanical Transport get into plenty of trouble and so do the E.O.s in the Engineers.

Oh yes! I'm officially known as a flight lieutenant. Have the small stars and a shoulder strap commonly known as a "Sam Browne." Furthermore I wear boots, and if desirous of so doing may wear gold spurs. Swell fellah!

'Bye, Izzy, and a world of love,
Bo

. . . .

January 26, 1918

Miss Isabelle G. Young
Kappa Alpha Theta House
Stanford University

Dearest

Here we are but very likely to be under way in a very few hours. We came aboard ship last night and are permitted to go ashore until one this afternoon.

I certainly was glad to have finished with railroads for a while. Ever since the first of the year I've been riding around on the darn things and my poor head is battered to a pulp by bumping it around in berths and dressing while gracefully poised on the small of my back and eating on dining cars and such like. Enough is enough.

Our boat is the Tunisian, an Allen Liner and not a regular transport. We are the first cabin passengers which is nothing to brag about as this tub isn't very large or particularly new. However it has been in service since the first of the war and only torpedoed once which speaks well of its commander.

There are about 2500 Chinese coolies—the ones we passed yesterday—in the steerage. That makes a very fair boat load.

I understand that we are going in a fleet with a couple of other ships and a convoy. If we do we ought to be at least a couple of weeks crossing as convoys are slow, particularly at this time of the year.

Incidentally my boat is No. 9—right upstairs on the boat deck—and the old life preserver is right above the bunk. This afternoon at four we have boat drill which is encouraging.

A Canadian infantry officer told me last night that one of the boats was taking over about 2000 Poles who are going to join the Polish army which is being organized and which is expected to reach 150,000 men. I'd heard the rumor before but this was the first definite news.

The part of town that was ruined by the explosion is an awful looking mess. There are blocks and blocks of ruins—nothing but piles of wood and brick where houses were. Trees have been stripped down to bare stumps. The docks are only rows of piles and boats are lining the shore.

The Imo—the Belgian Relief boat—was blown high and dry across the bay and the munitions ship was blown at least a mile away. Uptown all the windows were broken and plaster shaken down. It certainly is a terrible looking wreck and compares favorably with San Francisco in its best days.

Izzy, dear, you've no idea how I hate to go. It's not the going over that counts nor what we'll run up against when we get there but I'm getting so darn far away from you and maybe away for such a long time. I certainly do hope they speed up the mails a bit. If I hadn't been such a poor nut I'd have gotten home to see you and tell you all the things I don't seem to be able to write. But you know without my gushing about it don't you dear?

However it all has to be done and we might as well be real optimistic and pretend it's all fine and that we wouldn't have it any different if we could.

Send me some Dippies when you have time—any of that stuff that might be interesting. And don't forget a couple of new letters—the old ones are nearly worn out.

I must stop, dear, and get this ashore while I have a chance. Next one will probably be along in about a month and from "Blighty."

Love from

Bo

. . . .

[Seven letters from Liverpool Postmarked February 6, 1918]

January 28, 1918

Izzy Dear

Thank the Lord I didn't join the navy. We've been out only a day and many, many of the boys have lost interest in life already. It was really beautiful when we left port yesterday. The sky was of azure and the sea of sapphire blue—just like in songs 'n everything—and the shore was the pure white of the snow and the black of the woods. Very pretty and poetic. But this morning the sky was a sullen grey and the water a dirty green—also rough. I don't suppose it is really rough as North Atlantic storms go, but there has been a nasty north wind, and the port side of the boat is thick with frozen spray. Our fleet sticks pretty closely together and some of them don't seem to mind the weather while others take on a lot of water.

I imagine we'll all have plenty of it before we get to England.

Things are changing already. Tea is served every afternoon at four—a sliver of bread and a little cup of tea. Personally I can't see what particular good it does, but I guess it will have to be. "When in Rome etc." And when I asked for a slice of lemon yesterday the waiter nearly fainted. It

A view through a porthole of the Tunisian, *perhaps of the doomed* Tuscania, *mid-Atlantic, January 1918.*

seems that lemons are scarcer on the boat than coal in New York which you will admit is pretty scarce. . . .

January 29
. . . This craft has displayed an aptitude for rolling that would make any self—respecting billiard ball retire in blushing confusion. On top of that the wind and sea have increased and the barometer taken a drop. All very encouraging. Every time I think of ten days more of it I become firmly convinced that foul crimes will be committed and that the raving nuts will be legion 'ere we reach England. Already the best of friends are getting mean and saying nasty little things to each other.

January 30, 1918
The sea is rougher, the wind is stronger—altho not so cold—and the barometer has dropped. . . .
Time surely does hang heavy on the hands altho today has passed pretty quickly. This morning we amused ourselves by throwing cigarettes to the coolies. For one "Fatima" the boys will commit assault and battery, and for a dime they stop at nothing short of murder. At that they are a happy looking crowd and just like a bunch of kids.

In the afternoon I got tangled up in a small bridge game which was ended by boat drill at four. We enter the danger zone in a day or so and have to carry our belts with us all the time now. In the zone they put on an extra watch and take plenty of other precautions.

January 31, 1918

It's a perfectly grand day, child. Sixty mile head wind, wild and heavy sea, and a wonderful sun. For two hours I've been all over the boat and in doing so managed to get beautifully wet. The old tub buries her nose into a swell and then sticks it into the sky and shakes the water off. We went forward a while ago, and after nearly being washed away a couple of times gave it up as a bad job. There's a long lean American tanker next to us that is under water most of the time and looks more like a submarine than a steamer. It's great tho, to feel the salt spray in your face and the old boat plunging and pitching along under your feet. . . .

Got real chubby with some of the coolies this morning and one of them gave some yen in exchange for some cigarettes. The lieutenant in charge of them said it was very old—Manchu Dynasty I believe—but maybe he was kidding me. The stuff was probably made in 1908 or 1909. I'm a bit suspicious of the boys who tell you wonder tales. Maybe they'll bring you luck.

February 3, 1918

Today being Sunday the proper thing to do was to go to church which I did. Two Sundays in succession is something that hasn't happened for several years and it worries me. I feel religion coming on.

We are just about on the edge of the danger zone, and our gang are doing extra watch. They haven't reached the "Rs" yet but when they do Rogers will probably have the windy side of the deck and the graveyard watch. It never fails. There is a bonus of twenty pounds to anyone who spots a sub, which seems to keep the watchers spirits up to par.

The boat drills continue. The danger signal is a succession of short blasts on the whistle but as far as I'm concerned one solitary little toot will be enough to start me meandering in the general direction of the boat deck.

Seriously, tho, this submarine business is not particularly humorous when you come right down to brass tacks. These Huns are very much in earnest about the whole business and they pick off a boat every now and then.

From all indications we will dock Wednesday or Thursday which isn't such terribly bad time. However everyone will be glad to feel a little solid land under them again.

February 5, 1918

From reliable sources comes the glad tidings that one more day and good old mother earth will be under foot. There isn't any land visible to the nude optic as yet, but it's very misty and the fact that we turned south a couple of hours ago is encouraging.

The destroyers are with us having suddenly closed in from nowhere yesterday morning. Most ridiculous looking little boats they are bobbing about like skiffs and scarcely visible a mile or so away. But they're wicked and when they cut loose—oh la la la!! Nothing to look at but a cloud of smoke and two converging sheets of water. They have to be careful in the heavy seas as the big waves go right over them and they can't stand too much hammering. They impart a feeling of security that isn't at all unpleasant.

The weather has been very bad and consequently fairly safe as Gentle Fritz, the German sub commander, doesn't seem to be able to get in much dirty work unless it's fairly calm. Sunday night I had about as much chance of staying in my berth as I'd have of staying on "Cyclone." Last night wasn't much better. Yet it's much warmer in spite of the wind and I have hopes. . . .

The English monetary system has me flabbergasted and I'll probably fall for the British equivalent of "give me two tens for a five." Not only must one learn the simple arithmetic of the lucre, but also the many nicknames and mispronunciations. The most encouraging news to date is that six pence is a perfectly enormous tip, a soul-and-conscience-buying-fee, and that two and six (2/6) ($.60) is all that a soldier is permitted to spend for one meal. There's a chance for me yet.

I'll write as soon as we get to "Blighty" and tell you all about the place.
Considerable love from
Bo

. . . .

Imperial Hotel
Russell Square
London
February 8, 1918

Dearest

Here's your old friend Bo in Blighty and still fightin' em up. We arrived Wednesday night, reported to the Air Board yesterday, and are to be posted to squadrons next Thursday.*

[enclosure]

BIG AMERICAN TRANSPORT SUNK
Over 2,000 U.S. Troops Rescued
Liner torpedoed at night off Irish Coast
210 Lives Lost out of 2,397

Admiralty, Thursday.

The Anchor liner Tuscania (Captain J. L. Henderson) was torpedoed at night on February 5 off the Irish coast whilst carrying U.S. troops.

Total number on board 2397
Total number saved 2187

This is the first loss of United States troops at sea since the American Government started the despatch of their expeditionary force to Great Britain and France.

Mr. Baker, U.S. Secretary for War, stated a week ago that the United States would have 500,000 men in France early this year.

Only a fortnight ago the same Minister pointed out that the Germans were preparing a great submarine offensive against the American lines of communication with France, and said that the explanation of the recently marked decrease in the number of Allied merchantmen sunk by submarines was to be found in the withdrawal of submarines for the approaching thrust on land and sea.

The Tuscania, which has now been sunk, had a tonnage of 14,348 tons, was built in 1914-1915 on the Clyde, by Messrs. Alex. Stephen and sons, Linthous, and made her maiden voyage to New York in February of the latter year. . . .

*The censor erased Rogers's description of the sinking of the *Tuscania*, though he did nothing about an enclosed newspaper clipping on the subject from the *Evening Standard*.

They called for an extra watch, and I was one of the volunteers really preferring to be on deck. Until midnight I warmed a little strip on the port rail and nerves—every little wave developed into a torpedo coming directly toward us, every light was a sub coming up for aim, and the flashes from the lighthouses were all calcium spotlights featuring us. Talk about your dark alleys. Here we were sneaking along between Ireland and England, not a light showing. Night blacker than the ace of spades, and Gentle Fritz with all eyes and ears open. It's a funny, funny sensation and it surely woke me up to the fact that there is a war on. . . .

I suppose there will be a lot of hot headed and ignorant criticism in the states as to why the Tuscania wasn't properly protected. As far as I can figure out it has been a wonderful thing that this was the first boat to get it. The fact that only one ship was sunk speaks well of the navy. It's a wonder to me that they don't get many, many more. Anyway it was an exciting night, and a nice smooth harbor was a very agreeable sight the next morning.

We left for London late in the afternoon . . . and arrived shortly before midnight. . . . We took a lightless cab, drove up a lightless street, and stopped in front of a dark hotel. And here we've been since.

Dear, you've no idea of how evident the war is here. In just two day it's made itself apparent in countless ways.

Picture, if you can, Market Street with every young man in uniform, soldiers, sailors, and all. Add a large number of Australians, New Zealanders, Canadians, Americans. Remove from the street practically all pleasure cars and put in their places girls in khaki suits driving staff officers about in their own private cars. Add some convalescents, a few busses, and much bustle and hurry and you have the Strand.

Attempt a positive image of the St. Francis lobby with one half of it boarded off, a scanty sprinkling of officers and civilians, and all the fixtures covered. The Rose Room is filled with long wooden tables and busy girls; the Blue Room is partitioned off into offices; the Ladies Writing Room is a bureau of information; most of the rooms are filled with busy clerks, for the greater part women; and the Colonial Ball Room is a store room for records. That's what has happened to the Hotel Cecil, the famous Cecil. It's now the headquarters of the Air Board.

Walk down to breakfast tomorrow morning. Eat no fruit. Use no sugar—none. You may use a couple of saccharin tablets in your coffee, but nothing more. Go down to Palo Alto and buy some scrawny little oranges and pay fifteen cents a piece for them. Try to get some candy and find you can't. That's a sample of how things go here. There seems to be plenty of plain substantial food but no fancy things.

At the stations you see Tommies coming in covered with Flanders mud, rifles over their shoulders and iron hats strapped to their backs, and you realize that maybe less than twenty-four hours ago they were in the front line trenches.

I'll venture to say that you don't personally know of one person who intimately knows, or possibly has ever heard of anyone, who has lost his life yet. Over here, where at home they would have a service flag, there are honor rolls. Everyone has been hit and hit hard.

All this isn't pessimism, it isn't exaggeration, it's only cold facts. I had no idea of what a tremendous affair the war is, how terrible it all is, and how the English people have worked and sacrificed. I only hope and pray that the folks at home will wake up to the seriousness of the war, to the absolute necessity of sacrificing everything to win it. . . .

Enough heavy thot.

London, what I've seen of it, is very interesting. I've been banging around with the Harvard boy I told you about, Elliot Chapin, and we have a large sight seeing campaign mapped out. There are so many interesting and historical places to see that we scarcely know where to begin. Today we spent ordering and buying a lot of stuff we needed. I got a wiz of a trench coat for five pounds ($25), and a much better coat than you can get in the states for forty or fifty dollars. They are really quite necessary here as it rains a great deal. However the weather is very mild and pleasant.

I've only been short-changed once and that was because I forgot and said two pence instead of tuppence. I can talk fluently of bags, luggage, lifts, and the rest, and I'm beginning to like tea.

Love from
Bo

[The censorship apparently eased somewhat, because Bogart wrote his mother a letter that appeared in the *Los Angeles Examiner*.]

ROGERS, L.A. BOY TELLS HOW HE SAW SINKING OF TUSCANIA
Describes Wild Night of Terror as U-Boat
Strikes Vessel with U.S. Soldiers

An eyewitness to the sinking of the transport Tuscania in which nearly 200 American soldiers lost their lives, Bogart Rogers, son of Mrs. Belle Green Rogers, 1608 Las Palmas Avenue, has written a letter to his mother which vividly describes the sinking and the night of terror that followed.

Young Rogers, a well-known Los Angeles boy, is in the Canadian flying service and was enroute to England on another vessel, but in the same convoy as the Tuscania. He was on watch aboard the vessel on which he was traveling when he saw the explosion that wrecked the big transport.

HIS STORY

Here is the story as told to his mother in the letter which Mrs. Rogers has just received.

"The Tuscania was one of our convoy and at the particular moment it was hit, yours truly was on guard on the starboard bow and got a fine look at the whole business.

"At about 6:30 o'clock Tuesday evening I suddenly became aware of the fact that there was a war on. The Tuscania was about 400 yards to our starboard and about 100 yards astern. The torpedo came from the port side and therefore couldn't have missed us more than a few rods unless it was fired from behind the Tuscania.

DULL EXPLOSION

"There was a flash, a cloud of water and smoke, a dull explosion and the whole ship raised 10 or 12 feet, not out of the water, of course, but listed and then rolled back. That was all and that was enough.

"The rest of the convoy never even slowed up. The T.B.D.'s came right up. The Tuscania put on all lights and sent up rockets and we managed to increase our speed a trifle. If you think the rest of Tuesday night was pleasant, you're mistaken. I stood watch until midnight, when I was relieved.

"The night was black as coal. Every little wave developed into the white wake of torpedoes coming directly toward us, every light was a Hun submarine and every flash from the lighthouses was a spot light featuring us.

"It surely is a nasty feeling to know that Fritz is laying for you and not being able to see him.

QUIVERS WITH SPEED

"Our old boat quivered with speed until we reached port next morning. You are absolutely helpless at night, as there isn't a light. The report says that about 250 got it, which is quite a few. Anyway, that woke me up to the fact that something was happening on this part of the world. I've seen many, many signs of it since. . . ."

4. CHATTIS HILL *February–March 1918*

When he finished training in Texas, Bogart Rogers had accumulated twenty-nine hours of flying time in the Curtiss Jenny. In England, the contingent from Canada would be assigned to various airfields for further training in their assigned specialties, whether pursuit, observation, or bombing. After some three months' hiatus in training, pursuit trainees would regain their proficiency and sharpen their flying skills in the Avro 504, a two-seat biplane and the standard British trainer, which was lighter, faster, and more maneuverable than the Jenny.

Ultimately, the aspiring fighter pilots would make the transition into "real service scouts," primarily either Sopwith Camels or SE-5As. The latter, produced by the Royal Aircraft Factory, was rugged, dependable, stable, and maneuverable. The British manufactured almost 2700 SE-5As, a number exceeded only by the Avro. The SE-5 saw service in British, Australian, and American squadrons in France, Egypt, Palestine, Mesopotamia, and Salonika.

As Rogers continued his training, across the Channel on March 21, 1918, three German armies attacked along the British front with more than 1 million men. The anticipated great German offensive had begun. The Royal Flying Corps would be sorely tested and suffer heavy losses. Depleted squadrons would fill their ranks from the pool of pilots in the training squadrons.

. . . .

London
Imperial Hotel
Russell Square
February 11, 1918

Dear Child

... This morning and part of the afternoon we spent in Westminister Abbey and it's even more wonderful than I ever imagined. You can wander about for hours looking at the tombs of celebrities who date so far back as 600 A.D. and there are memorials to just about everyone who has ever been really famous in England. One of the most interesting things is the Wolff Memorial upon which almost all the colors of Canadian regiments in France have been placed. Then there are all the old kings and queens and the famous statesmen and poets and soldiers and the old cloisters that are crumbling away with age and the beautiful windows and a hundred other things.

... Tomorrow we're going to see the opening of Parliament and try to get a general idea of St. Paul's and the Tower and Tower Bridge.

Yesterday and Saturday we saw the Royal Academy, Parliament, and a lot of the river. There is a war exhibit at the Academy that beats anything I've ever seen both in photographs and every sort of relic. There are some German machines that have been shot down and all kinds of captured articles, and the pictures tell whole histories.

The busses are my chief joy and I could ride around on top of one of them almost all day. They go everywhere and slowly enough so that you can see what passes, and the fare is never more than tuppence.

A good many machines are run on gas as a substitute for gasoline and they carry huge canvas tanks on top giving them the appearance of young Zeps. . . .

Tomorrow night we've booked a couple of stall [front orchestra] seats for the Gaiety where a large musical spectacle called "The Beauty Spot" is on. London theaters seem to be having a big season and I've heard that a lot of good shows are running.

The food really isn't so bad when you find out where to go for it and what to order. Officers are only permitted to spend five and six (about $1.25) for supper so you see we can't go broke buying meals. However you can get about anything you may wish for that. Fruit and sweets aren't too plentiful but there's enough of everything else. . . .

I guess the Tuscania affair has aroused a lot of feeling at home from what the papers over here say. I certainly hope it has, as people can't wake up any too soon. I'm no pessimist, but I can't possibly see any decisive end to this war until the United States gets a great many men in the field and makes some great sacrifices. Say what you will it's a nasty fight and one that must be won no matter what the cost may be.

Wholesale love from
Bo

. . . .

Imperial Hotel
Russell Square
London
February 15, 1918

Dearest

This must of necessity be a short letter for in a very few minutes we grab a taxi, dash to Waterloo Station and depart for Chattis Hill Aerodrome, Stockbridge, or maybe it's just the other way around. Anyway it's the camp I've been posted to, and as the squadrons there are equipped for scouts it looks as if yours truly was due to be a scout pilot. As you may know they are little single seater fighting machines, fast as the dickens and capable of doing almost anything if you know how to handle them. I'm perfectly satisfied, as there are a lot of advantages in not having an observer to bother with nor be bothered by. You're your own boss and don't have to worry about anyone but yourself.

Last night I had a long talk with an instructor from one of the camps, and he told me things about new types and methods that it was hard to believe. It puts the old imagination out of whack trying to realize them.

I can't seem to find out much about Stockbridge aside from the fact that it is about seventy miles from here in the southeast and is a comparatively new camp. As there are no regular quarters at present, we will probably be billeted.

Chapin is going to a camp up near Liverpool, which makes sad parting. I'm going this afternoon with a Beta from Vanderbilt, a chap named [Evander] Shapard who is exceedingly southern in speech and manner, but who is a pretty nice boy. Somehow or other I seem to run to Betas.

Most of the A1 sights have been seen. Tuesday we saw the opening of Parliament and got a glimpse of the King and Queen, of the Prince of Wales who is a darn nice looking boy, of General [John Denton] French and Admiral [John Rushworth] Jellicoe* and a lot of others.

Then we saw St. Paul's, which is beautiful and the tower, which reeks of history and the crown jewels and all the old armour and Sir Walter Raleigh's cell and St. John's Chapel and the place where Queen Anne was executed and the room in which the two little princes were smothered and the Bloody Tower and Traitor's Gate and—anyway we saw everything around the Tower. Then we went down to the docks and the Tower Bridge

---

*General French was the first commander in chief of the British Expeditionary Force; Admiral Jellicoe was the commander in chief of the Royal Navy.

and up to Threadneedle Street and the Bank and out thru Whitechapel. But sightseeing gets tiresome, and I'll be glad to get back to work.

I've had the most terrible time trying to keep from getting run over. The traffic all goes the wrong way, and as one of the first lessons of my early childhood was to look first to the left when crossing a street, it's rather hard to change. And at night the busses and taxis carry almost no lights, and it's impossible to see them and they can't see you. Great place.

Kid Shapard just blew in and informed me that if we had any hope of catching our train we had better get under way. So good bye, dear child, and more from the wilds of Chattis Hill.

Love
Bo

. . . .

Stockbridge
February 17, 1918

My Dear

We left London Friday and after a couple of hours and a change of trains arrived. The ride was a most interesting one and I got my first good look at rural England. It's all it's supposed to be and reminds you of someone's front yard.

This particular town is in Hampshire, about seventy miles from London, and right in the center of the English stock farm district. In fact, but later. It is in a small valley and on the banks of the River Test, a babbling young stream. The town hall, one of the newest buildings in the town, was put up in 1810 and the stone bridge over the Test dates 1799. Using those as a basis the rest of the town was here in about 1600. The houses are mostly of brick or plaster supported by oak beams and the roofs are thatched or of red tile covered with moss. All of the houses are warped and twisted and look everyday as old as they are.

Eight of us are billeted about a mile from town at The Grange, a certain Lord Herdon's fishing cottage. It's a very comfortable place, plenty of room, a large fireplace downstairs, feather beds, and all the modern inconveniences such as no running water, lamps and candles, a trick bathtub and refrigerator temperature everywhere except in the one room. But we have an old housekeeper and a batman [orderly] who takes care of all of our stuff so it really isn't bad. We have about a mile to walk to mess, which is in the hotel in town and a couple of more miles to the aerodrome but one can usually get a tender or truck out there. The walk in the morning gives you an enormous appetite and when we walk home after dinner we are ready for another meal.

This evening I walked back from supper with this chap [Shapard] I told you about in my last letter. . . . I suddenly discovered that he is engaged to a Theta at Vanderbilt. A small world I repeat. Anyway we talked about lots of things and let the old moon get the best of us. . . .

Golly, Izzy, I'd give everything I've got or ever expect to have to be in Palo Alto tonight wandering along up by the lake or up on the hill with you. If the mail doesn't show up shortly I'm going to go nutty.

The aerodrome here isn't bad and the machines are wonders—wicked looking little planes with enormous roaring and rumbling engines and the stunts these pilots do with them—I thot I could imagine what scout flying would be like but I guess my imagination must have been out of whack for it was way ahead of expectations.

These little busses will get up terrific speed and then loop and roll and spin in the craziest way you can imagine. I don't see how I'm ever going to learn to fly one, but others have so there's some hope for me.

What I started to say about the stock farms was that one of the boys here was hauled up before the colonel tonight for diving on horses. It seems that he'd been amusing himself by diving on the stock around here and making them scatter. However he picked out a swell looking farm and scared a regular race horse that was being primed for some large meeting next week. The horse ran into a fence and got all tangled up, and the owner made a kick. However the boys must do something for excitement.

All my love to you, dear, and more, both love and letters, shortly.
Bo

. . . .

> Royal Flying Corps
> Chattis Hill
> Stockbridge
> Hampshire
> February 22, 1918

Dearest Isabelle

So this is Washington's Birthday? Funny but they don't seem to make any fuss about it over here. We had an argument at lunch today with a couple of English boys and one of them pipes up "Who was this fellow Washington?" Whereupon young Mr. Shapard of Tennessee says "Who was this guy Nelson? I hear a lot of talk about him over here."

Anyway we spent most of the day in jumpers working on engines. It dirties the hands, face, and clothing, but brightens the brain. It's not a bad idea to know something about motors and the best way to learn is to tear 'em down and look at the insides.

It has been raining all day, not hard, but a steady drizzle that has soaked everything and made a sea of mud of the aerodrome. There was no flying which accounts for the manual labor.

Yesterday afternoon I finally got into the air, the first time since I left Texas. It certainly seemed like home to be floating around up there with most of England beneath, the channel and the Isle of Wight to the south, the Test twisting along in a little green valley, and the rest a checkerboard of green and brown fields.

My instructor, a chap named Gibbons, goes by the nickname of Reckless Reggie and he's all of that. His idea of a good time is to take off over the hangars, shut off the motor, and then yell and wave his arms at his friends on the ground. He sits in front and has a telephone thru which he gives instructions to me in the back seat. The poor nut sings songs and talks to himself when he isn't talking to me. When we did a loop he kept mumbling "Here we go. Up Reggie, up, up. Over she goes. Oh! Pretty loop, Reggie. Pretty loop." And he does tight spirals and dives on people and carries on in an alarming manner. I'll be darn glad when I go solo and won't have him to worry about. . . .

I've been reading lots and lots of things since we came over and I'm really getting interested again. From Ruskin to O. Henry is a far cry. . . . Ruskin, Dumas, Balzac, Bret Harte, O. Henry, and all the boys. . . .

Love, and lots of it, from
Bo

. . . .

Chattis Hill
Stockbridge
February 25, 1918

'Lo Sweetie

. . . Most of the mornings are used up by flying or working on machines and engines, and in the afternoon there is more flying and an hour of gunnery and another of wireless or photography and after tea—yes, we can get tea at five, if we happen to feel a bit weak or in need of a stimulant there is always a lecture on some subject or another. This evening it was on Aerial Fighting.

The day was pepped up a bit by having an instructor and his pupil fall thru one of the new buildings that is being built. Their engine "conked

Lieutenant "Reckless" Reggie Gibbons, Rogers's instructor at Chattis Hill, with a cane and teapot.

out" and they turned down wind with disastrous results. The machine went thru the roof of the building, deposited the pilot and his seat on the floor, and scattered engine, gas tank, and such all over the place. Neither of them were badly hurt, a broken ankle and nose and numerous cuts and bruises being the sum total. Anyway it was excitement.

Good night
Bo

. . . .

> Royal Flying Corp
> Chattis Hill
> Stockbridge
> March 3, 1918

My Dear

Here it is Sunday evening and absolutely nothing has happened since Saturday evening. So goes it in Merrie England. Last night it rained. . . . Everything was "washed out" for the day, and after lunch came back out here and here I've been ever since playing bridge part of the time and plowing thru "Les Misèrables" the remainder.

I managed to get quite a bit of reading done, and for want of anything else struggle with all the stuff that you should be able to say you've read. Just finished "Lord Jim" by Conrad and for the first time managed to get interested in his stuff. . . . Must get educated and highbrow and everything. . . .

About the British officers you saw at Wilson's—if one of them had a green stripe on his sleeve, and it was near the shoulder it was a regimental designation. The different battalions in the field wear all sorts of distinctive marks, usually either on the sleeve near the shoulder or in the middle of the back right under the collar. It may be a square or a circle or a triangle or almost any figure and in different colors.

Wound stripes are a gold bar worn on the left sleeve at the wrist. You are entitled to one for every visit to a base hospital and it's not uncommon to see even as many as four over here. . . .

How about Ellen Calhoun? You haven't mentioned her for letters and letters. I can't get over thinking about Pop now and then altho it's terrible the way time covers up such things. I suppose it's best after all, as the way troubles pile up these days you'd go nutty if some of them didn't fade into the past.

'Bye, lover
Bo

. . . .

                                                      Royal Flying Corp
                                                      Chattis Hill
                                                      Stockbridge
                                                      March 5, 1918

Dear Child

Last night I started to write to your mother [about their engagement] and lost my nerve.

Oh well! Just as long as Isabelle knows and cares we can manage to fix things up somewhere and somehow.

Today was another "dud" one. No flying, no nothing but a couple of classes. It refuses to rain and get it over with, but stray drops fall intermittently all day long.

But yesterday we had a grand time, me and Reggie. We went up and first did forced landings. To enlighten you we would fly away over the country, and suddenly Reggie would shut off the motor, point to some little two by four field and I would proceed to get into it minus a motor.

Then we tried cross wind landings which are just that and nothing more. Next came a bit of instruction in side slipping, which is a grand way of getting down in a hurry. After that we chased other machines around and I learned a few of the methods of attacking another machine, diving on them, coming in from a blind spot, or getting underneath their tail.

As a grand finale we staged a swell vertical dive which ended up going about 120 per straight for Mrs. Earth. If everything goes nicely I may get on service machines in a week or so.

Tomorrow being our squadron holiday I'm going to crawl out very, very early and run into London for some clothes and a square meal and a hair cut and lots of things. . . .

This is all, lover. Must run up and pack my grip.

Wholesale love from

Bo

. . . .

                                                      Royal Flying Corp
                                                      Chattis Hill
                                                      Stockbridge
                                                      March 7, 1918

Dearest Isabelle

. . . This morning we got to talking over the possibilities of the Calgarian, which was torpedoed a few days ago carrying a lot of mail. Believe me, Fritz can sink boats full of food and boats full of men, and boats full of munitions, but when he sinks boats full of mail from home it's high time to step forth and fight.

Yesterday I rode up with a major who had been at Mons and the Marne and almost everywhere in France, and the nasty things that man could find to say about Germans in general was a joy and a blessing. He was most interesting and able to speak from experience. He was quite optimistic and of the opinion that America would be able to swing the balance quite quickly once her armies were in the field.

Today was another "dud" one and there was almost no flying. In the morning it was foggy, and in the afternoon a thick, dirty ground haze put the kibosh on things if I may be permitted to use the vernacular of the hoi polloi. . . .

Good night, lover
Bo

. . . .

Chattis Hill
Stockbridge
March 9, 1918

Dear Child

There's no news to speak of except that yesterday afternoon I went up in a heavy haze and was lost for half an hour—hadn't the least idea where I was and couldn't see except directly below. I finally picked up the River Test and flew up and down until I located the village and the rest was easy.

This morning we had the first clear weather in several days and I went up for my photography. Most of the country for miles around was visible, and after taking a couple of towns near Stockbridge I went over the River Avon and down across Salisbury Plain to Stonehenge. The whole plain is one vast camp or series of camps, infantry, artillery, and half a dozen aerodromes. There are trenches and artillery emplacements and in certain places the ground is a mass of shell and bomb craters. From there I went to Old Sarum which is an old Roman fortification a couple of miles north of Salisbury. It's a great round hill surrounded by a moat affair and circled by walls right to the top. At Salisbury the cathedral is the most conspicuous mark. From there I followed a broad white road back to Chattis Hill.

The first casualty since we came happened this evening when a chap in our squadron either had his bus catch fire in the air or crashed and then burned. Anyway it was a nasty mess from what the fellows who saw it said.

It's sort of depressing to have such things happen, especially when you know the fellow who gets done in but I guess it's all part of the game.

The RFC and the Royal Naval Air Service are being combined and are to be called the Royal Air Forces. The uniform is to be different—I'll tell you all about it in the next letter—but most of the fellows are going to stick to what they've got. New uniforms are expensive luxuries these days.

Wholesale love from
Bo

. . . .

>Royal Flying Corp
>Chattis Hill
>Stockbridge
>March 10, 1918

Izzy Dear

Today was a grand day, and this afternoon I indulged in a little mild aviation taking an aeroplane up to 11,000 ft., the highest I'd ever been. Barring the fact that it's cold as the dickens up there, it's really nicer than it is down below. In descending you have to swallow or hold your nostrils and blow to relieve the pressure on your ears. Otherwise you'll have a wicked headache.

Tomorrow or the next day I may go on service machines—machines which are actually in use at the front which are small and fast and hard to handle until you know how.

Yes! Yes! Terrible news! Heartrending news! News that would make strong men weep and women curse! There is a rumor—only a rumor, of course—that much, much mail went down with the Calgarian, but most of it from Canada. I understand that most of the American mail comes directly across on American boats so maybe it won't effect me. Torpedoing mail is the nth degree of meanness.

Good night sweet child
Bo

. . . .

Royal Flying Corps
Chattis Hill
Stockbridge
March 13, 1918

... The camp starts moving Friday, one squadron going then, another Sunday, and our gang on Tuesday.

This evening's doings were wild and hilarious, a very fair dinner, much liquor, music, and promiscuous waving of the arms. Also we drank to the King which is quite the thing to do in a British officer's mess. When the food is all disposed of the president rises, glass in hand and says, "Mr. Vice, the King." Whereupon everyone else rises and Mr. Vice says "Gentlemen, the King." That's the cue for the gentlemen to murmur "the King" and drink hearty. Very impressive. For further particulars see Kipling, "The Man Who Was." In the good old days it was customary to snap the shank of your glass but there's a war on and glasses are hard to get.

I've been getting in a lot of flying during the last two or three days and am getting so that I can throw an aeroplane around with abandon and some judgment. Monday I spent the whole afternoon above the clouds. They were only at about 1500 feet and it was perfectly clear above them. Things directly underneath were visible thru holes and rifts but there was no radius of vision. Yesterday morning I went cross-country, first to Andover, then to Old Sarum. Landed at both places, had my papers signed, and made some calls on a couple of fellows I knew there. In the afternoon I fooled around for a while and had a "scrap" with another machine. Believe me, a practice fight sort of gives you an idea of what a real one may be like. You're absolutely alone, can't hear a thing except the thunder of your own motor, and you never look at the ground or notice where you may be drifting to.

All the time you just have to keep your eye on the other fellow and twist and turn trying to keep him off of your tail and at the same time try to get on his. However it's a lot of fun, especially when you know that he can't do you any harm if he does get in position for a sitting shot at you. This afternoon I just fooled around some more, but my wind shield broke and I had a fierce time with oil flying back in my eyes. As goggles are an awful bother and rather obscure your vision most of the fellows depend on the windshield. The type of motors we are using now are rotary, that is, they revolve around a fixed center. Centrifugal force throws the oil, heavy, dirty, sticky castor oil, out of the ends of the cylinders and the wind drives it back in your face. A big drop of oil sort of lengthens out like a rubber band and hits you "slap." Very uncomfortable when you get it in the eye. Also very dirty and fatal to the clothing.

March 14

... I shall write on and on until one of the gang here proceeds to start making toast and producing from our larder a jar of rather doubtful jam. None of us can get a good night's rest any more unless we agitate our internal workings a bit beforehand....

... Cox & Co. ... are—ahem—my agents and are very good about handling almost anything. They are located just about opposite the Admiralty and less than a block below Trafalgar Square. Incidentally English banking methods are not all that American methods are. For instance instead of making up your pass book and returning the checks they write every item out in the book....

Do you remember how I used to babble about India and what a grand place it was. And how grand I used to feel when you worried. But this is all different, Izzy. If I ever get back I'll never leave you, dear, not for India or any other place, never, never, never.

And now some more tiresome junk about flying.

This morning one fellow landed right on top of another one who was taking off, and for a minute things looked bad but they were both only about fifteen feet up and managed to land in one piece, altho the machines were busted up considerably. About two minutes later another chap tried to make a fancy landing in front of the hangars and knocked over a large pile of bricks which were to be used on a new building. Result: a broken machine and much profane language from his major.

About noon a couple of scouts from another camp appeared from a cloud and dived at 200 per on the hangars. One was painted a bright orange with red trimmings and black stripes around the fuselage. It contained a pilot who was a wizard and nothing else. His first stunt was to dive until he was about three feet from the ground and going over 200 per and then zoom straight up to about 300 ft. where he would loop and come out about a hundred feet up....

After that he rolled and spun and went into loops, got to the top of them, turned out one way, and then instead of going on over simply did a half roll and was all right side up and pretty. Later when he came down, we discovered that he was a captain with twenty-nine Huns to his credit, Military Cross, and a D.S.O. Incidentally he was flying the same type of machine that we are going over on.

After that everyone seemed to go crazy, and all afternoon machines were flopping about and doing funny stunts. That sort of thing is contagious....

'Night, sweetie
Bo

. . . .

Royal Flying Corps
Chattis Hill
Stockbridge
March 16, 1918

Good evening, Sweetie,

This is just about a deserted village. 91 Squadron flew away yesterday morning. 92 leaves tomorrow, and I will swill my Sunday evening tea in solitude as the one remaining comrade goes with them. We dash away Tuesday. We all hated to leave this place for it has very comfortable and almost homelike. From all reports Tangmere—the new camp—is nothing to look forward to.

I was talking to a chap this morning who had just transferred to the RFC from the infantry who said he received a letter a week or so ago that was mailed from Boston in March 1916. It had followed him all around France and England and finally caught up with him. Army mail may not be fast but apparently it's sure....

Bo

. . . .

Tangmere, Sussex
March 20, 1918

Dearest

Suddenly out of the cold and rain of yesterday Spring has come. The trees and shrubs are turning a most delightful green—a dozen shades— and at night when the warm winds blow off of the channel all the pink cheeked country lassies wander thru the village. The gay warble of the bluebird mingles with the honk of the motor truck, the smell of fresh green things growing is richly associated with the odor of benzol, petro, gaso, and a dozen other "lines."...

We've been working all morning putting up hangars as the camp is only about half finished. The hangars come with a pile of girders, several rolls of canvas, and a box of bolts and metal fittings. Getting them together is like working out a jig saw puzzle. However when they are up they make fine, large sheds.

We have a fine large mess hall with lounging rooms and electric lights and our quarters aren't at all bad. But both are only about half finished, and the plaster is still damp.

The aerodrome is large and level as a tennis court and covered with turf like any front lawn. It's situated in a very pretty part of Sussex, about five miles from the Channel and mid way between Portsmouth and Bristol. The nearest town of any size is Chichester which is about four miles away. Tangmere is merely a hamlet situated back of the camp.

The night before we left Stockbridge one of the squadrons that is taking our place there came in with several fellows I knew very well. . . .

. . . [O]ne of the fellows flying over here yesterday crashed at Romsey, a town about ten miles from Stockbridge and who should he see there but Harry Jackson and everyone of the bunch I chased around with in Texas. They had been there for a month and I've flown over that camp time and again. Here they were only ten miles away and I'd been writing to them c/o Postmaster, N.Y. Jack sent a note, and I'm going to fly over and have lunch with them in a day or two. It's only about forty miles from here, and I'm nutty to see them. The world is small and England is smaller.

You're right about all the people training, but if you realized what kind of a game modern warfare is you'd understand why so much training is necessary. Take for instance flying—scouts have to fight, and in order to stand a show in a scrap you must be an absolute master of your machine. Loops, and rolls, and spins, and any number of fancy turns and stunts. Not only do you have to know how to do them, but you must be able to do them almost instinctively as you don't have a chance to watch your machine, especially when you run up against a circus. Shooting is largely a matter of luck and the only sure way of getting your man is to get right on top of him—ten or fifteen yards—and the only way to get where you can shoot him is to be able to outfly him. But at the same time you must know enough about your gun to fix stoppages in a hurry—in seconds. All of that takes time and practice and you're absolutely no good if you aren't an expert. The same is true of every branch. Men simply have to learn a new trade—or should we say profession.

Courage and gameness aren't worth a row of pins unless the knowledge is with them. So you see, lover, it's an enormous job teaching millions of men how to fight effectively.

As to when we may go over—nobody knows. It may be soon, it may be months. We may go as a squadron or be posted to squadrons already in France. I know not and care less—in fact going over would be a welcome change. And it's not as bad over there as you might think. Quite a nice place in fact.

These particular barracks are cursed with two phonographs or gramophones, as they are called over here, and one or the other is going incessantly sometimes both at once.

More love than usual from,
Bo

. . . .

Chichester
Tangmere, Sussex
March 22, 1918

Dear Lover

You nearly had a letter written to you last night but your old friend Bo was a sick child—not exactly sick but very, very low. The cause of this indisposition was a ball game between 91 and 93 Squadrons, and for want of a better I pitched for 93. Needless to say we were neatly trimmed 6 to 4. Everything was rosy until the seventh inning when I lost the range and this, coupled with an error or two, did the work. Last night the whole carcass felt very much as if someone had been walking on it, and there is only a slight improvement this morning. . . .

Active work will start again about Monday and we probably won't have so much time from then on as the days are getting longer and we can fly about twelve hours per. However we are to have no classes and that should help a little.

You asked what kind of machines we fly, and if I haven't told you before—they are single seater scouts, very small and very fast, the best of them going close to 130 an hour. As they have to be taken off and landed not less than seventy miles an hour, you have to do it just about so or they are up on their nose or over on their back and nicely crashed up. They will go just about as high as you care to take them, and believe me it's not warm over 10,000 ft. They mount two machine guns and about a dozen instruments—altimeter, air speed indicator, compass and a lot of others. The cockpit is very small and all cluttered up with these instruments and the controls, and you can just about get into it. In fact, if you wear too much bulky clothing you're going to have trouble.

There are any number of different types of machines for different purposes ranging from tiny scouts to enormous bombers but the little ones are certainly the prettiest.

We also have larger machines here, two seaters for dual instruction, but they aren't service machines and aren't used overseas.

Love
Bo

. . . .

Chichester
March 23, 1918

Izzy Dear

... Things are looking up very nicely here at camp. The mess is getting better every day and we are going to have a tennis court, a bridge tournament, some kind of a concert, and lots of wild things. As soon as the place begins to get comfortable I'll probably go to France or someplace.

At present everyone is very much interested in the German offensive which has been under way only a couple of days. The general opinion seems to be that minor gains will be the only result altho it's pretty early to tell just what will happen. Maybe it will wake people at home up to the fact that this is a wicked war and that winning it is going to be a tough proposition. But the darn thing must be won and won completely no matter what the cost and just between you and me the cost is likely to be considerable. Strangely enough people in London are no longer able to appreciate moonlight nights. They make the place too visible from the air. But they don't have air raids in Palo Alto and—Oh! What's the use.

Good night, lover
Bo

. . . .

Chichester
Tangmere
March 26, 1918

Dearest Isabelle

Primarily, it has come, the sweater. And Izzy dear, the odor has ruined my young life just as I knew it would. You should have "iced" it in the coal bin or packed it in tobacco but as it is—anyway I'm not going to wear it until the smell has entirely disappeared.

Today was another wonder and for amusement this morning I flew way out over the channel which looks very pretty especially early in the morning. France wasn't visible as there was a heavy bank of fog and clouds a few miles out. However I'll see it soon enough.

Now they say that a confession is good for the soul so in case you think that we are worked to a state of exhaustion[,] here is the daily program as set down at this particular camp. Three mornings a week we are forced out at 6 A.M. After tea and toast we start flying at 6:30. Breakfast at eight, or whenever you may not be flying usually takes an hour or more. Then more flying till one when we have lunch. The afternoon is our own altho there may be classes later. Tea is at four, a lecture at 6:30 and supper at 7:45. The other three days we have the mornings off and fly from noon until dark. We have enough batmen so that our only worry is to keep them supplied with shoe polish and find things for them to clean. Two or three hours flying a day is plenty and as much as four does you up nicely.

Good night, lover
Bo

. . . .

Tangmere
March 28, 1918

Dear Lover

... Today has been perfectly "dud" as far as flying is concerned and the only thing on the program was a lecture on compass flying this afternoon.

Yesterday afternoon by the grace of God and the skin of my teeth I became a graduated service pilot by the simple process of flying a real service scout [SE-5A] around the place, taking it off, riding it around and landing it without busting any of the essential working portions, smearing up the landscape, or breaking my neck. They are wonderful little machines, Izzy, heavy but they fly like a bird and go to beat the band. The first time I started to land I sailed in over the edge of the aerodrome and thot maybe I was going a bit fast so looked at the air speed indicator. One hundred and twenty five miles per hour it said, whereupon I decided that maybe we should go around again and come in a bit slower. The next time we got on nicely at about eighty. But the way the things are built and the smoothness with which they fly makes the sensation of speed almost nil.

Sunday being Easter I'm seriously considering going to church. There can't be a more appropriate day for such a move, and it's really quite the thing to do. Besides, lover, somehow or other this affair makes you think of religion more than one might think.

As a little chap who sleeps next to me says—"This is a long war and a wicked war and a damn reliable war." So 'tis, so 'tis.

Good bye, dear
Bo

5. TANGMERE, SCOTLAND, AND FRANCE
April 1918

On April 1, 1918, the Royal Flying Corps, the Royal Naval Air Service, and the Australian Flying Corps collectively became the Royal Air Force. Bogart Rogers thus became a charter member of the Royal Air Force. Yet the wings he wore bore the letters *RFC*, and he would always identify himself with the Royal Flying Corps.

On the Continent, the British, their backs to the wall, as General Douglas Haig, commander of the British Expeditionary Force, warned his troops, struggled to stem the great German offensive, which had torn a large gap in their lines. Although 1232 RFC aircraft confronted 1000 German planes on the western front, the Germans outnumbered the British at the point of attack: 750 to 580 aircraft. In the air, the RFC quickly nullified the German air force's impact on the battle, regained numerical superiority, and proceeded to win aerial ascendancy over the battlefield. Its aggressive strafing attacks harassed and ultimately helped to stem the German advance. One fighter plane flew so low that it ran over a German company commander.

Yet the cost was high. During a ten-day period in April, the RAF lost 478 airplanes, more than had been lost during the previous two months combined; by April, it had lost 1302 airplanes. The RAF sorely needed pilots.

. . . .

Tangmere, Sussex
March 29, 1918

Izzy Dear

Things proceed as usual in this part of the world, and everyone proceeds to do as little work as possible. I've got a grand batman who is making me lazier than ever. He's a regular old maid, tends to everything and does lots of worrying for me. He tends to laundry and polishes shoes, belts and buttons and puts my sleeping bag out in the sunshine—when there happens to be any—and if I leave a letter on my shelf, he gets a stamp and posts it for me. I'm getting perfectly useless.

However I managed to get in an hour and a half this afternoon on the service scout and had a great time. They will fly absolutely hands off for any length of time. All you have to do is sit back and look at the scenery. But the landing was awful, and I nearly ruined myself and the machine.... But as Reggie [Gibbons] might say "That was a bloody bum landing." Tomorrow morning before breakfast we practice many, many landings and will strive to do better. This one was too close for comfort.

After the first of the month the Royal Flying Corps will cease to exist. We are being consolidated with the Royal Naval Air Service and the Australian Flying Corps into the Royal Air Force. I don't imagine it will make much difference in the way things are handled as the main object is apparently to get all flying forces under one head.

There is to be a change in uniforms which will not become effective until after the war but then French blue is to be official. Personally I hate the stuff. Khaki is alright but when they get to brilliant and passionate blue I fail to enthuse.

Wholesale love from,
Bo

. . . .

Chichester
April 2, 1918

My Dear

As usual excitement is rather scarce. The formation we waited for on Sunday finally got under way yesterday morning—after a fashion. One machine crashed taking off, another broke an aileron wire in the air, and the third nearly blew up owing to a leaky radiator. I flew rings around Chichester waiting for them to show up and finally landed and called it a day. This morning I had an hour's fighting practice with my instructor and if he'd had a gun I'd be dead now. I'll be darned if I could keep him off my tail and I couldn't get on his. Then I went out over the channel and chased a "blimp" that was patrolling the coast.

Tomorrow I'm going on graduation leave; probably a couple of days in London and a day or so at Romsey with Harry Jackson....

The daylight saving days are working nicely. Altho it is nearly nine now there is still a faint streak of light in the west. It is possible to fly until seven thirty and they say that later in the summer it's perfectly light at nine. All of which is very well as long as I'm here and you're there, but were you here I'd say it would be rather annoying.

Did I tell you that this place was between Portsmouth and Bristol? My mistake. It's between Brighton and Portsmouth and almost directly south of London. A mere trifle, but it's always nice to be precise. . . .

More love than usual,
Bo

. . . .

National Hotel
Upper Bedford Place
Russell Square
London
April 3, 1918

Dear Izzy

Dear Old Blighty!!! Foggy and drizzle and cold and dark. If there happens to be a copy of Service's "Rhymes of a Red Cross Man" around the place look up the poem about going back to Blighty. Kid [Evander] Shapard quotes the last line

"To sniff the air of Blighty in
the mawnin' "*

and then laughs sarcastically.

However it's not such a bad place after all and quite exciting after Tangmere.

There are three of us here, Shapard, Wilson, and myself. We are planning on seeing a few of the sights that we missed before and getting everything we need as it is undoubtedly our last chance. I don't mind telling you that I shouldn't be at all surprised if all three of us were in France inside of two weeks. I only hope we will go over at the same time and be able to stay together.

More tomorrow, dear, and until then just a little love from,
Bo

. . . .

---

*Robert Service, "Going Home."

London
April 4, 1918

Dear Izzy

... This afternoon we went to see the exhibition of war pictures at the Grafton Galleries. They are all photographs in color, many of them life size, and a more wonderful lot of pictures it's impossible to imagine. Some of the details, particularly in facial expressions, are beyond description.

There's a story going the rounds here about the King's recent visit to the front. He was visiting some wounded Canadians when one of them said to him, "I've heard a lot about you. Glad to see you. Put it there." And the King, being a regular person, "put it there."

There is a noticeable scarcity of men and officers about town, which shows that there is work to be done in France. I feel perfectly ready, even anxious, to go. While there is much to be learned about flying and fighting, there is no better place to learn it than France.

You inquired what I do for amusement. The last time I indulged in any aviation was enlivened by practicing several loops, some rolls and half rolls followed by climbing turns, an Immelman turn or two, and then a couple of spins. They all have to be practiced continually and also such things as a half loop and a half roll which brings you out right side up. ...

I can't draw an Immelman or a spin but in the first you go to the top of a loop, kick on opposite rudder, and go down the same way you went up, coming out in the opposite direction from which you started.

A spin is produced by simply stalling your motor and kicking on rudder and aileron. Your nose falls, and you spin down with your nose as an axis and the machine spinning around it. And believe me, they spin fast.

I've probably bored you to death with technicalities so enough. I love you, Isabelle, and I'm going to get back to you even if I have to be so rude as to do in a couple of Huns to do it.

Good night, dear
Bo

. . . .

National Hotel
Russell Square
London
April 5, 1918

Dearest
I done it. It's all over. Just finished a letter to your mother.
I'm just scared to death that maybe your family don't approve of me—

April 1918 83

which wouldn't be at all strange — and I can't write them and tell them what a nice boy I am and try to boost the Rogers stock. But I hope more sincerely than I ever hoped before that they don't object too strenuously. . . .

Much, much love from

Bo

. . . .

London
April 6, 1918

Dear Child

My ship has come — filled with mail. I'll be busy with answers for days and days. . . .

Also sent you a copy of Bairnsfather's "Fragments."* You may not be able to appreciate all of them but, they are really wonderful cartoons and immensely popular over here. Some of his best ones — those of the Italian front and with the American Army — aren't in book form yet. Shap sent one to his lady. We remarked how nice they would be to show Sunday evening callers and then both cursed in unison. The one about the same moon may appeal to you. The one of Cox's is nothing but the truth.

Tomorrow we go back to Tangmere and I'm not sorry. One soon gets fed up on London especially when the thots and desires are thousands of miles away.

Bo

. . . .

Tangmere
April 8, 1918

You 'Ole Sweet Thing

. . . Yesterday morning we went to St. Paul's — getting alarmingly religious — and it is really a wonderful place on Sunday. We were coming back here in the afternoon but couldn't get a train until evening and finally pulled into Chichester rather late, grabbed a taxi — it was pouring — back to the old place again. Unfortunately they had a bad day yesterday, three fellows going west in one crash — a collision in the air. It was the first fatal accident we've had here. There were also several minor crashes. . . .

Nothing definite has developed about our going away yet. We may get a chance to go to a special fighting and gunnery school in Scotland for a

---

*Bruce Bairnsfather was to the Tommies and doughboys of World War I what Bill Mauldin was to the GIs of World War II, a cartoonist with the common touch.

week or so. Nobody knows. Here we have far more pupils than serviceable machines so they can't get us under way too soon for me.

The thing I hate about this game is that after you've been at it for a while you sort of get fed up on ordinary stuff. Getting overconfident you tend to become careless. Once you get careless you're gone, absolute gone. Regardless of what happens you've got to keep the old head up.

Love,
Bo

P.S. I'm just about to be a First Lieutenant.

. . . .

Tangmere
April 11, 1918

Dear Child

The weather: cloudy and misty. Very little flying. Rogers flew an Avro for forty minutes, practicing forced landings in nearby fields, thereby amusing the natives and stampeding the stock. Swell afternoon.

Izzy dear, I'm completely fed up on this place and they can't ship us away any too soon. I guess we will get a week or so in Scotland at a special fighting school. Any old place will be better than this.

Had two packages from mother today—both candy. Real American candy that is made of sugar and not flour is a treat. Bring it on!!

Good night, Isabelle. Don't you forget about
Bo
(Lt. Bogart Rogers RFC)

. . . .

Tangmere
April 12, 1918

Miss Isabelle G. Young
Kappa Alpha Theta House

Just between you and me if I felt any better than I do right now, I'd be arrested or shot at dawn or some such thing. On a day like this it's simply impossible to be otherwise. Rogers arose rather late, spent an hour on the range, then got hold of an old Avro for a little joy ride. Izzy dear, there aren't adjectives enough in the dictionary to describe that ride. First I headed west toward Southampton climbing all the time. Managed to get up to 12,000 over Portsmouth and then turned out over the channel. It wasn't even cold up there, dear, and to the west the sky and sea just merged in an expanse of hazy blue-greys and grey-greens. To the south France was just visible, and it's over a hundred miles across here. Below was the Isle of Wight with its great white cliffs and the little harbor of Cowes. Down the coast I could see Beachey Head sticking its nose into the channel and nearer Brighton and Shoreham and Littlehampton.

There wasn't a ripple on the water. It shimmered like silk in the sun. There were several T.B.D. [torpedo boat destroyer] patrols and a couple of "blimps" scooting around off Bognor apparently looking for a sub, for they chased around in circles leaving pure white streaks in their wake. The Hun hasn't much of a chance in weather like today's. From the air every contour of the ocean bottom is visible, and a sub is easily seen by the blimps.

Honestly, Izzy there are times when floating around up in the air is absolutely beyond any description. Later about noon I led a formation around the place for about an hour. There's no such thing as too much flying on a day like this. But there are few machines and many pupils, so this afternoon I borrowed a bicycle and peddled merrily down to Chichester, looked over the crowded streets, and peddled back again. Then played around the hangars for awhile swinging propellers and such. Then a glorious cold shower and dressed up all pretty'n everything. And now I'm writing and waiting for supper—nearly famished. This is the life—in a way, and what's more I saw the new moon over my right shoulder last night.

I think—think only—that we will be leaving for Ayr, Scotland, the first of the week for a course in fighting. It's a grand course, and we are pretty lucky to be able to go there, especially now. It will also be something to see Scotland, or at least to get a glimpse of it. Might as well see the whole place while we're over here for—God willing—if I ever get across to the right side of the ocean, I'm going to stay there. . . .

It's eight, lover, and I'm starved. Providing the gas doesn't get me you're likely to hear from me again.

I love you, Izzy, honest.
Bo
(Lt. Bogart Rogers, RFC)

. . . .

Tangmere
Easter Sunday
April 14, 1918

Dear Child

Having seen my duty and done it noble I'm conscience free and feel sweet and pure 'n everything. Or, in other words, I went to church this morning with a chap named Wilson and—but why not make a short story long.

Four of us were scheduled for a formation at 9 this morning. We were going to fly down to Brighton and entertain the Easter crowd with a little playing around. It was pretty windy, but we rolled the machines out, tied streamers on the leader and decided upon what positions we should take. Then we warmed up our engines and, in spite of a very gusty west wind, were all ready to start when the major showed up and decided it was too breezy for a formation. And so it was. We staked the machines down and came back to our stove. The machines are still lined up on the aerodrome waiting for the wind to go down, but it's hopeless. The zephyrs increase.

Then it was that Wilson and I felt religion coming on. We put on decent looking tunics and brushed our hair all pretty and left on our filthy dirty breeches and pu[t]tees and hied ourselves over to the Tangmere church. . . .

Honestly, Izzy, this was the most—well, there was more feeling of real religion there than in any other church I've ever seen. It is a little stone building with a tall steeple and surrounded by a neat little yard filled with graves. Some of the stones are worn absolutely smooth—scarcely a mark to be seen—and many of the others date back before 1800. The church itself is of stone and brick. On the inside great hewn oak rafters support the roof. The walls are just plain white. Above the altar is one stained window. This morning it was so simply decorated, no wonderful Easter lilies or hot house flowers, but a few yellow jonquils and some calla lilies, sort of unpretentious and tasteful. And the old ladies in mourning who wept when they prayed and the hard boiled little choir boys dressed up to look like little angels. You see, dear, you go to church so often and I so seldom that the things that seem quite commonplace to you make quite an impression on me. And at the memorial church the women don't weep yet.

After we came back a machine bound for Gosport got lost and in trying to land on our aerodrome turned over and busted up much of the external workings. I have the tip of his broken propeller and tomorrow shall attempt to make you a picture frame. Several of the fellows have made some very snappy ones; at any rate they are something of a curiosity. It may take some time to make, but you shall have the results. . . .

Bo

. . . .

London
April 14, 1918

Izzy Dear

It's a long, long story, but at any rate here we are in London again headed for Scotland and very shortly for France. Last evening about six the adjutant hunted two of us down, Wilson and myself, and calmly told us to get our stuff together, pay our mess bills, get a transfer card from the quartermaster, see the medical officer, and then report to him, as we were to leave for Ayr at 9:30. Great excitement. We dashed around madly, couldn't find the M.O., nor the Quartermaster, nor the mess secretary, had to send my batman out for some laundry, finally got everything fixed up, all the possessions together, and fixed everything up with the squadron, so we are all ready for France. We took a terrible train from Chichester, had to change at Brighton and then—we had checked all our luggage—get that word—to Victoria Station, but the train we were on went to London Bridge and we were supposed to change. Being almost one A.M. when we came to our changing point, we had all gone to sleep. Our luggage went to Victoria; we landed at London Bridge. It was well after one, raining, and as the station is across the river a long way from any place, we couldn't find a taxi or a bus or nothin'. Finally a YMCA car came to the rescue and took us up to this place. The YMCA has risen to a high place in my estimation.

Now on our orders it said we were to catch a train at St. Pancras Station at 8:50 A.M. for Ayr. We matched to see who would crawl out in the cold dawn and get the baggage from Victoria to St. Pancras. Needless to say I was the odd man. Very humorous.

Now in London Sunday morning is devoted to sleep and late breakfasts and such. The tubes don't start running until nine, no breakfast before that hour, and the busses get on the job around ten. I managed to get some tea and toast for breakfast and wandered out in search of a taxi. Deserted, Isabelle you never saw such a deserted place in your life. I walked from Russell Square to Piccadilly Circus, a good mile and a half. The only living creatures I saw were two policemen, one truck driver, three dogs, and a cat. No taxis, no busses, no cabs, and as it was after eight I began to worry a bit. Finally I found a taxi and dashed away to Victoria. We loaded enough stuff for three trucks into this old, one cylinder hackney carriage, and chugged back to St. Pancras. Wilson and Shapard—oh yes—Shap came with us—were not yet there, so I inquired a bit. Strangely enough nobody seemed to know of a morning train for Scotland. There was the "Flying Scotchman" at 8:50 P.M., but no train in the morning. I cursed. Then we all came back here, had a large breakfast and here I am, writing to you lover.

I don't expect we'll be long at Ayr, a few days at the most and then overseas. And believe me, I'm not sorry. Last night there was a small draft of infantry at the station, iron hats, guns, and full kit. It's something to know that you aren't going over that way....

Yesterday morning we went down to Shoreham and were gassed, but no casualties. First we were fitted with masks, then instructed in the use of them, and finally put thru first tear gas and then chlorine. While you feel absolutely no effects from it, I'd surely hate to have to work or fight with a mask on for any length of time. They aren't very comfortable and it must be pretty disagreeable work. There are people—jealous no doubt—who inferred that I was better to look at wearing a mask than without one.

The next—from Scotland.

My dearest love, Isabelle,

Bo

. . . .

Middleton, Ayr
April 16, 1918

Dear Child

"Here we are at Killey, and a bonnie fine town it was before the war," said the Highland captain. "Weel, weel," remarks Shapard. "Wirra, wirra," says Wilson apparently under the impression he was in Ireland. "Hoot mon," says I and we proceeded to climb off the "Flying Scotchman" and onto the Scotch soil this morning.

At Kilmarnock we got a train for Ayr, and after reporting to the station O.C. [officer commanding] were assigned to billets. At present everything is rosy and—but we're ahead of the story.

Primarily it was a shame we had to come up on a night train, for we missed a lot of perfectly well behaved scenery....

At Kilmarnock some nice YMCA ladies fed us hot tea and some sort of mysterious Scotch cakes. Another boost for the YMCA.

Ayr is quite a desirable town. Get out the geography and you'll find it on the southeast shore of the Firth of Clyde. It's fairly large—some 35,000—and very clean and prosperous. The population, however, doesn't run true to form. I almost expected to see them all in kilts and wearing Tam o Shanters, but they dress quite like other mortals and strangely enough talk so that they can be understood. It might be well to add that if

Scotchmen wear kilts here, I blush to think what they'd wear in California. For it is rather chilly and damp. But the lights shine at night; we get real butter; and there are a few almost mountains. Really, Izzy, it's way ahead of London and south England. . . .

We have a very nice billet, an enormous old private house only a block from the water and surrounded by very pretty grounds. The mess is very nice and is run by WAAC [Women's Auxiliary Army Corps].

However what pleased me most is that right across the street is a golf course, and this very afternoon I played eighteen holes. This is surely the land of golf. We saw some grand courses coming up. Prestwick, Turnberry, and Troom all being more or less famous.

The aerodrome is an old race track and not too good. But the machines make up for all other deficiencies, for they are in good condition and almost every type even including some captured German machines. The instructors have all had a great deal of experience, and most of them are decorated from V.C.s [Victoria Cross, Britain's highest award for valor] on down. One pilot in particular is now the ace of aces and has over fifty Huns to his credit and just about all the decorations in the category.

Just how long we may be around is a gamble. Some stay two weeks, others two days. I'd like to get in some flying here but it doesn't make much difference one way or the other. . . .

Good night dear lover,
Bo
(*Lt.* Bogart Rogers RFC)

. . . .

Middleton, Ayr
April 17, 1918

Dearest

Three days in Scotland and golf every day which is at least as it should be. We are in the officer's pool waiting to be posted to squadrons, but I imagine we'll be flying in a day or so. It happens that just at present there is little demand for pilots flying the type of machine I've been flying—which is encouraging—and therefore we aren't being worked very hard. However any day a call may come and out we'll go.

This a great place, Izzy. Has it over England in dozens of ways. and I'm enjoying myself hugely. In the mornings we have to report for orders. At noon there is a lecture on fighting by Captain [James] McCudden, V.C., D.S.O., M.C., M.M., D.C.M., Mons Star, and Croix de Guerre. He is the kingpin of British pilots at present and a very nice chap, quite young and not a bit swelled up by his honors.

We have to report again at two thirty. After that nothing to do 'til tomorrow.

It seems that this part of the country was the stamping grounds of our old friend Bobby Bur-r-r-rns. There are a couple of statues of him about the town. A car runs out to the Burns cottage, and the people seem to be quite proud of him.

The place is really quite Scotch, little kids running around in kilts, and people with broad Scotch accents, and Scotch toffee, and Scotch plaids, and even Scotch whiskey. There seems to be plenty of everything which wasn't so in England.

Yesterday afternoon we walked down the water front and took a look at some of the small shipyards. The larger ones are all farther up the Clyde near Glasgow but this whole section is the center of the shipping industry.

In the evening everyone parades along the promenade for this is quite a popular summer resort. There is a long walk along the ocean front, a pavilion, a large park, moving picture shows, 'n everything. Just at present it is light at nine. . . .

Much, much love,
Bo

. . . .

Middleton, Ayr
April 19, 1918

Dear Lover

. . . McCudden gave a wonderful talk this morning—sort of opened up a bit—and made it very clear that successful pilots are so only because they have worked like sin, studied every phase and detail of flying, machines, and the habits and haunts of the Hun. To hear him talk nonchalantly of doing in Germans at 20,000 feet and of studying all available material in order that he may know where to go to look for them convinces you that this is surely the greatest game God ever created. There's nothing like it.

Part of our bunch has been sent away to Turnberry. Shapard and I are the only two remaining. However, it's only a matter of time before we are pried apart.

Love from
Bo

. . . .

Middleton, Ayr
April 21, 1918

Dearest Isabelle

. . . Still hanging around doing nothing and getting well fed up on it. They don't seem to have a very great demand for my particular brand of pilot, but I would like to be able to do a little flying. This afternoon my roommate was going to take me out to tea, but it was rather cloudy so we didn't go. That is one of the favorite pastimes up here. Two fellows get into a bus, fly away over the countryside, and locate a large and prosperous looking house with a convenient field. They then proceed to land in the field, climb out, and begin to tinker with the motor as tho it were the cause of all the trouble. Whereupon the simple country folk—not always as simple as you might think—come out to look the daring birdmen over. You get real clubby with them, and generally an invitation to tea ensues. When it's over you have someone to swing your propeller—that's the object of two going—and fly away home.

How would you like to have me come buzzing along, land on the Theta front lawn, and have Sunday night supper with you? Not half as much as I'd like to. Someday maybe we'll be making Sunday afternoon calls on friends and relatives in such a fashion. You never can tell what will happen in this day and age.

This morning we walked out to Burns's Cottage and looked around the place. I daresay it's much cleaner and better kept than when Bobby lived there, for the paint has been liberally applied and what was probably at one time the barn yard is now a carefully kept garden with lawn and shrubs and neat little paths.

Then we went to Alloway Kirk which is about a mile from the cottage and only an ivy-covered ruin. There are gravestones in the yard that date to 1691, and the cornerstone of the church very indistinctly says 1516.

On the "bonnie banks o'Doon" is the memorial. As it was closed to visitors today we went around and crawled over the stone wall. Then there is the "Auld Brig o' Doon" and I must say that the Doon is a stream that might inspire a wooden man to beautiful poetry.

The whole place is organized to get the unsuspecting tourist and must have done a thriving business before the war. . . .

Almost supper time, Izzy dear, and you know all about me and supper.
Wholesale love from,
Bo

. . . .
Middleton, Ayr
April 23, 1918

Dear Lover

The daily communiqué has nothing to report. We show up every morning and afternoon, the adjutant says nothing doing, and I silently curse and play golf to pass the time away.

However this morning there was a little show staged on the aerodrome that may be worth writing about. Two fellows started up in a Bristol Fighter, one as pilot and the other in the observer's seat. In taking off the pilot did a steep climbing turn down wind, which is generally a foolish thing to do. The result was that the machine lost all flying speed, side slipped and crashed right in the middle of the aerodrome. A good crash usually makes an awful noise and this was no exception. The bus lay there for a second or so, and every one from the hangars rushed out toward it. Then—poof—and the whole thing was in flames. The observer had a broken arm and was sort of half hanging from his cockpit. When the fire started he crawled out in a hurry his clothes burning nicely. He rolled over on the ground, put the fire on his clothes out, and then went right back into the flames and dragged the pilot, who was unconscious, out and rolled him over. It all happened so quickly that it's hard to explain, but the observer surely showed wonderful presence of mind and a considerable amount of nerve. Both of them were rather badly burned, but very fortunate in getting out at all. You've no idea, Izzy dear, how quickly a machine will catch fire and how completely. The gasoline does it. All that remained after a few moments was a few blackened metal parts.

That's the one thing everyone fears—fire. Crashing or getting shot down isn't so bad, but being penned in a machine perfectly conscious and yet unable to get out is ghastly. And catching fire in the air is even worse.

Who should pop in today but Bill Taylor from somewhere in north Scotland and full of news and gossip. And the wretch said "Bo" you're getting fat." But not really fat, Izzy, just less—slender than usual. Heaven forfend that I should ever get fat.

Good night, dear.
Bo

. . . .

Royal Pavilion Hotel
Folkestone
April 27, 1918

Dearest Isabelle

In a very few hours your lover will be in [that] dear France and darn glad of it. We came down from London this morning and are leaving sometime this afternoon. . . .

There was a call for pilots at Ayr, and here I am. We left there at night, arrived in London yesterday morning, and left this morning so you see it required much hustling to get a great many little things accomplished.

London seems to be about the same which is neither a boost or a knock. . . . [T]here was a raid warning, maroons and whistles. Many flocked to the shelters and tubes but apparently the raiders turned back for nothing happened.

This is all for now, Izzy dear. Bill Taylor and I are going shopping some more. We haven't an awful lot of time. I'll write volumes from France.

Love from
Bo

. . . .

France
April 28, 1918

Dearest

So this is Paris!!! Oo la la!! But it doesn't happen to be Paris at all but only a God forsaken depot [Berck] some little distance from the front and a more forlorn looking spot can't be imagined. We finally arrived here about two A.M. after a ride on a French train that would make some of the much vaunted Peninsula trains—the 10:35 or the 4:40 for instance—look like lightening expresses. It was raining beautifully, and we had to report and find billets and put up our beds and finally—but I'm miles ahead of my story.

We crossed from Folkestone to Boulogne on the smoothest sea and the clearest day one could desire. We were convoyed by T.B.D.s and seaplanes but nothing exciting happened. At Boulogne things began to happen.

First the luggage. It is removed from the boat in a large rope net. Said net is suspended some fifty feet it seemed above the dock. Suddenly all the stuff is dumped out. Rather strenuous treatment for trunks and suit cases. Then a porter inquired in perfect English if we wanted a porter. We did and thot "aha, here is a friend indeed, a man who can speak English." But he was deceiving us. The one phrase was all he knew. However he spoke

Rogers (with pipe) on his first crossing of the Channel, April 27, 1918.

The quay at Boulogne.

voluble French ably assisted by a Spencerian freehand movement and a suggestion of the Australian crawl stroke. After much precise gesturing we got it thru his head that our stuff was to go to the gare Central. Assisted by English military police we got our stuff checked and started out in search of food. . . .

We found a respectable looking hotel with a waiter who could converse in terms of food, but who was unable to compre any financial turns. Anyway we had a grand steak and omelet even if it nearly broke the bank paying for it. The steak may have been horse, but the eggs were bona fide. . . .

As we were leaving a Red Cross train pulled in direct from the front and filled with nice fresh cases. More rude awakenings.

We had very little idea just where we were going. At every station we would collar a guard and repeat the name of our destination. He would shake his head and unloose a rapid fire string of French verbs, nouns, adjectives, and possibly a swear word or two all seasoned with many ultra expressive gestures. But we finally arrived. The sooner we leave the better.

This depot is merely a pilots' pool where we stay until posted to a squadron.

I've finally discovered where all the white bread and butter is. It's here in France. Also much of the sugar, for we have unlimited quantities at mess.

I'm going to try to send you my ring, dear, and possibly it can be done from here. In England they would neither register it nor insure it for America, but it may go from here.

Things are looking up, and there ought to be plenty to write about from now henceforth and forevermore. Amen.

Considerable love from

Bo

6. MOUNT KEMMEL *May 1918*

At the end of April, a German advance in Flanders was just ending. Seven German divisions had driven French and British units off the high ground of Mount Kemmel. The Battle of the Lys had punched a thirty-mile-long by twelve-mile-wide bulge in the front line. In early May, the RAF counted its losses after Lys and issued orders for replacement pilots.

. . . .

[Berck]*
On Active Service
April 30, 1918

Dear Child

Still in the same place and likely to be here for several days, as the weather has been dud for nearly a week, and there have been few postings. We do a lot of range work every day which is fine practice. The old eye improves steadily. We shoot away a lot of perfectly good ammunition but the king pays the bills so it's nothing in my young life.

This morning it rained so we listened to a few lectures. We went out to lunch—five of us—to a little place up the line run by madame and Peggi and little Garcon. It was something to write home about. Peggi is a dreamy eyed, languid looking French girl who speaks English after a fashion and plays ragtime on a piano that apparently has never been tuned. Madame is large and dark and smiles always and shrugs her shoulders frequently. We had a real salad, plenty of French bread and butter, an omelette, a steak, French fried potatoes and peas, coffee and nuts, and a bottle of rare old red ink. The whole works was very well cooked and miles better than anything I ever had to eat in England.

Now for a few answers to the letters. First about mail from here. Offi-

---

*Locations in brackets added by editor. In France, censorship rules prohibited the naming of locations.

cers censor their own mail by the simple process of signing their name in [the] lower left hand corner. But of course we are supposed to be acquainted with all the rules and restrictions regarding mail from the field and have to act accordingly. However its nice to know that with very few exceptions your mail won't be opened. Just how much longer mail will take from here I don't know, but probably very little as the postal service is supposed to be good.

By now the letter I wrote [to Oregon] should just about have arrived and I'm going to start another one very shortly. I'm sure your mother will answer. You see, Izzy, I don't really know your mother and father as well as I might, and this furnishes a splendid opportunity for working up a speaking acquaintance. There might be a lot of things going on over here that would wildly interest them. I know your father is more or less of a golf nut—just how bad a one I'm not sure—but there's an opening that might be used to advantage, ranting about Prestwick, Troon, etc. a bit.

From all present indications my new spring hat is going to consist of one ultra warm helmet fabricated by the nimble fingers of a certain Miss Young. It will be useful, Izzy. The SE5 merchants—SE5's are the machine I'm on—do most of their work well above 15,000 feet and when you are up there for two hours at a stretch it gets a bit chilly. A good knitted helmet under a leather flying helmet helps matters out a great deal. . . .

Lover, there are times when I get to wondering when this mess is going to end and how. Sometimes the corners of the mouth droop a trifle and the spirit drops a few degrees, but it isn't as hard as you might think, especially when I think of all the grand future that's ahead. I'm quite positive now, that I'm coming back.

There have been times when things didn't seem so rosy, but a letter or so from you would pull me out. I'm completely cured and quite confident. All of which does not mean that you should stop writing.

Goodnight.

Bo

. . . .

[Berck]
On Active Service
May 1, 1918

Dear Lover

The sun seems to have gone to some other part of the globe for we haven't seen it since landing over here. This afternoon Bill [Taylor] and I went in to town and looked the place over. Not much to see.

Oh yes!! About knitting socks. That's the very best little idea you've had in many and many a day. Us troops must have woolen socks—the feet being all important—and the home knit ones are ultra, ultra. For Izzy dear, when one is flying the pedal extremities are invariably the first part of the carcass to freeze up. You have to keep them on the rudder bar, perfectly still, and they are down where very little heat from the engine goes.

Bo

. . . .

[Beauvois]
On Active Service
32 Squadron,
May 2, 1918

Dearest Izzy

By golly there must be a war going on from the sound of things. Just after supper tonight there was a terrible uproar down the line a way—seems that Fritzie was dropping a few bombs on the country side and the archies [antiaircraft guns] were after him. Quite a new sensation for your old lover, but nobody else seemed to mind it in the least. However we're way ahead of the story again.

This morning I was posted to the 32nd Squadron and early in the afternoon a tender called and toted Rogers and all his chattels away to war. We had quite a pleasant ride out here—a perfect afternoon and country that surely looks like California. It was hard to believe that a wicked war was under way—everything seemed so peaceful, men and women working in the fields, fruit trees blossoming and budding[,] and people carrying on in the villages just as if nothing had happened. But presently a string of great armoured cars rattled by, then dispatch riders, dust covered and hunched low on their motorcycles, became more numerous. The farther we went the thicker became the procession, big trucks loaded with men and supplies, little trucks, tenders, motorcycles and side cars all going or coming and all in a tremendous hurry. Strangely enough there wasn't a horse in the whole procession. This surely is a war of machines, machines and men.

We finally turned off the main road. After a mile or so across the fields we pulled up in a dirty little village—muddy streets, ramshackle houses and buildings, chickens, dogs, and children running about, trucks and tenders parked along the roadside. At a dingy little building on a corner with "Au Trocadero" in large red letters over the door we stopped. That was the officers mess.

Being just in time for tea I must stop and have a dash.

After being shown to a billet and unpacking all the worldly possessions I went to the aerodrome and reported.

That's about all for the day. There wasn't much there to see that hadn't been seen before, except as we were leaving a patrol returned from the lines.

If first impressions go for anything it's a pretty fair place. The O.C. is very popular, the fellows seem to be far above the ordinary, the mess is very good—at least supper was—and if Fritz wouldn't come so close with his bombs, I think everything would be lovely.

There are thousands of things to say, dear, but I'll say a few every day and stop for tonight.

Much love from,
Bo

. . . .

[Beauvois]
32 Squadron, France
May 5, 1918

Dear Lover

To begin with there is a steady stream of troops passing down the road in front of this place. As our window is only a yard or so from the road we can hear them quite plainly—artillery they are for the most part—the guns and caissons rumble and rattle along thru the mud, the drivers swear softly at the horses who slip and stumble thru the slime, but everything seems quite orderly, and they make rapid progress. Where they are from and where they are going I haven't the least idea and couldn't say if I did.

The weather has been very bad, heavy rains today and yesterday and low clouds and mist which have prevented flying. There's practically nothing to do on days like this—we get up late and simply loaf around trying to pass the time. Even when the weather is good it's an idle life at the best. Some times there are two patrols a day, others only one. If the patrol happens to be an early one, you're up early—otherwise you're not. When the show is over you're thru for the day. Of course there is your machine and guns to be cared for, but you have two mechanics who know every detail of the machine and armourers to take care of the guns. The pilot's job is simply seeing that everything is properly done. Both machines and guns are kept right up to the highest notch at all times.

Last night I had a grand time. There is an old major of engineers who is in charge of some work nearby. He eats at our mess. He's a typical old soldier—been in the army all his life, India, South Africa, almost every place on the globe, knows all about the army, is small, rather fat—or probably portly is the better word—partially bald, red-faced, grey moustache, and wears a monocle. For some strange reason—he says it's because I'm the only American in the squadron—he's taken a fancy to me, and last night we had a great pow-wow. He smoked cigars, put away many whiskey and sodas—which was unfortunate for every time he drank one I had to drink a small glass of port with him—and with every drink he became a bit more talkative. He babbled of Punjabs and Fuzzy Wuzzies and the Lancers and the Huzzars and Africa and Mandalay and bridge and horses and about almost every other subject under the sun. Of course I couldn't be outdone by a major even if he was old enough to be my grandfather, so he heard all about what a swell place California was; what wonderful girls one might find there, which interested him immensely; why I thot it advisable to double a one bid on suit; and why flying officers didn't deserve any more pay than the poor devils in the trenches. Oh, he got an earful before we were thru. He's a great old chap, altho undoubtedly he'd be an awful bore if given half a chance. He's as full of interesting tales as an infantry man's shirt is of cooties. Very inelegant simile that last, but quite picturesque don't you think, Izzy dear?

The war must be under way this evening, for every once in a while the windows rattle a bit and the sky lights with flashes. But it's a long distance from here and doesn't affect the sleep a particle.

Was dragged into some perfectly strange game of chance this evening absolutely against my will. That was one of the seven times I intended writing. Just out of pure meanness I took away seventeen large francs. What will the boy do with all this money!! Looks like thrift stamps or a Liberty Bond for me.

Oh yes. A chap got three Huns Friday inside of an hour. The last one he drove down intact far inside of our line, landed beside him, and came home with many souvenirs. Quite a fair show for one patrol.

Still have many, many things to tell you, lover, but a long time in which to do so.

Wholesale love from
Bo

. . . .

[Beauvois]
32 Squadron RAF,
France
May 7, 1918

Izzy Dear

For the last hour and a half I've been censoring the men's mail and am completely fed up on letters for the evening and maybe a lot longer. . . .

The reason for all this censoring is because Lt. Rogers is squadron orderly officer for the day and has to be on the job from early morning until early the next morning and inspect the men's quarters and their mess and inspect the guard and sleep out here at the aerodrome in a tent and tend to things in general. Thank goodness this job comes only about twice a month.

From the sound of things the war must be doing well this evening, for the guns have been at it hard and heavy for several hours. I wish they'd lay off—it's very annoying.

Bonne nuit, lover,
Bo

. . . .

[Beauvois]
32 Squadron RAF,
France
May 8, 1918

Dearest

. . . This morning the squadron went out on a patrol, ran into an awful mess of Huns, and had a horrible dog fight. All of our machines returned, some of them rather full of holes, but there are several Huns who didn't and who never will. There was another show this afternoon, but nothing happened. I surely will be glad when my first trip over is scheduled.

There's a chap here, a young Scotchman, who has sailed around the Pacific most of his life and who knows San Francisco, Portland, etc. like a book. We had a reminiscence party and talked of many things.

The guns are hard at it again tonight, and the windows rattle incessantly. A large push is coming off in a day or so—it was scheduled for today—and this must be a sort of prelude. Fritz is surely full of fight and now is his only hope of getting a decision. It looks as if the Germans were willing to sacrifice almost anything to break thru now, but it can't be done—bend probably but never break.

Lots and lots of love, Isabelle
Bo

. . . .

[Beauvois]
32 Squadron, France
May 10, 1918

Dearest

As for the new Air Force uniform it's an awful looking affair, and no one out here is getting it. The old major I told you about saw one the other night and said "My word, this gentleman looks like a cross between a commissionaire and a P and O boat steward."

Starting out with the very much overworked ego, I've been working on my machine part of the day. I cut away part of the fuselage in order to have a better vision over the side, had the stop pawl of my Lewis gun reset as the magazine was continually overriding and giving double feeds, all of which is doubtless as clear as mud to you. I moved the starting magneto back where it wouldn't stick in my side, put a wire fastener over the Very's light rack, as all the lights fell out the other day when I did a sloppy loop, had a catch on the engine cowling fixed, as it was loose, camouflaged all the white portions of the machine, said portions being the numbers and the emblems which were likely to attract attention, put in a new air speed indicator, and that was about all. This afternoon I studied a lot of maps—every pilot has five or six different ones—but couldn't get an awful lot from them.

Funny thing happened yesterday morning. A large bomber returning from a raid ran out of oil and had to make a forced landing. He had one bomb in his rack and was afraid to land with it, which was very foolish as they can't explode while in the rack. He cut the pill loose at random and it fell near a lot of French troops who were working on a road. Fortunately none of them were hurt, but their Captain came up to the aerodrome waving his arms and sputtering. For a few minutes it looked as if there might be an inter-ally war.

Would you like to hear about a patrol? Very well. The orders come thru from Headquarters, some times late at night, again early in the morning. This morning we got orders to stand by from nine thirty, all of which means every machine and pilot must be ready at any time. Nothing happened until this afternoon. About three the orders came thru. "You are to meet such and such a squadron who have been strafing Hun aerodromes

"A" flight, ready for take-off.

over Blank at 18,000 at five o'clock and bring them home." The order goes out to the pilots that they are to leave the ground at four, to the flight sergeants in charge of the machines that they are to be started and the motors warmed up at a quarter to four. Then everyone wanders down to the mess for a cup of tea.

About three thirty all the machines are wheeled out and lined up in front of the hangars. They really look most business like with their dark grey paint and their guns mounted and their general ferocious appearance.

The flight commanders carry streamers, usually one from each strut and different colors for different flights. The deputy leaders carry one streamer on their rudder. At a quarter to four the engines are started, twenty, 200 horse power motors just ticking over; twenty shiny propellers glistening in the sun; twenty—my mistake, forty—exhausts rumbling and belching blue smoke. Then the pilots straggle up and prepare for the journey up above. Preparing consists of putting on light, but very warm fur lined suits, sort of jumper affairs with fur collars, lined sheepskin flying boots, fur lined helmets, and polishing up the goggles. They have a little confab, the commander telling them what's to be done, looking over their maps and straightening things out generally. Then they climb into the machines, get nicely strapped in, maps arranged, gloves on, and everything fixed just so. Always before leaving the ground the pilot runs up his motor. Blocks are in front of the wheels, a mechanic at each wing, and one hanging on the tail. The motor is run up, slowly at first, then to top speed the

"A" flight, Thirty-second Squadron.

machine trembling and straining at the blocks. Everything ready, the pilot waves his hand, the blocks are pulled and the flights taxi out to take off, one flight at a time. Away goes the commander, motor roaring, streamers flying, and the rest follow in their proper formation order. A couple of turns around the aerodrome to get a close formation and they're away to the line—up, up and they soon disappear in the haze.

That's the start.

A couple of hours later a machine will come sailing down the road only thirty or forty feet up and running wide open. The mechanics can spot their own machines almost as far as they can see them—know by the way they are flying or by the markings or the color—and they run out to meet them. A turn or two around the field to see where the wind is, and they land—lightly as a bird—and taxi back to the hangar. Usually the first back is an old hand, for they know the way home and seldom get lost. Then they straggle in twos and threes and singly until the whole flock is in the roost. The fellows are generally pretty tired, especially if there's been a scrap, eyes blood shot and as nervous as a bunch of hopheads. Invariably the first thing they do is light a cigarette. Then they start comparing notes, and how the conversation does fly. "Did you see those three Huns east of ——?" "Why didn't you follow me down when I dived at that two-seater?" And so on 'till the conversation ends, usually sometime after dinner.

All of which probably means nothing in your young life. You see, Izzy dear, I'm so crazy about this stuff that I think everyone else is, which they probably aren't.

Archie worries the patrols a great deal, but really does little harm. The Huns shove it up all around machines but almost never hit them. Archie—if you don't know—is anti-aircraft shrapnel and the Huns are very clever at it. They can get alarmingly close even at 18,000 or 20,000 feet. Often machines return full of holes from it. However direct hits are very scarce, praise be.

You've been annoyed enough for one evening, lover. Another time you shall hear about A Flight and who is in it and all about them.

Good night, dear, and lots and lots and lots of love from,
Bo

. . . .

[Beauvois]
32 Squadron, France
May 12, 1918

Dearest Isabelle

Having spent a very busy day playing bridge I'm about to enjoy myself for a while—for my chief joy these days is writing to you, dear.

The weather being more or less bad all day there was no flying—nor was there yesterday. As the squadron possesses the distinction of being a sort of "circus" we apparently only work when there is something exciting on.

Now about A flight. The flight commander is an Englishman [S. P. Simpson], a big rather good-looking chap who has had lots of experience and who wears a lavender silk handkerchief in his sleeve. The deputy leader is a New Zealand boy, full of pep and one of the nicest chaps in the squadron. The chief Hun getter of the bunch is an Irishman from Belfast [Walter "Bing" Tyrrell] who is a picture of Joe Braden and a wild man if there ever was one. According to the French system of aces—a man who has five Huns—this boy has been one for some time. However they don't play up aces in the RFC. Next comes a Scot from Glasgow ["Scottie" Hendrie] who is full of Scotch wisdom and who has a very nasty tongue. Number five is a Canadian [Jerry Flynn] from Victoria, and last—Rogers. But they are a fine bunch of fellows, Isabelle, most of them are well under twenty-five, and the kind you feel willing to take a chance with any place knowing, if you get in trouble, they'll be right on hand to help you out. The other flights are just about as good. There's a great deal in feeling that you're working with fellows who think of you first and themselves

Captain S. P. "Simmie" Simpson, Rogers's first "A" flight commander.

Captain Walter A. "Bing" Tyrrell, a fifteen-victory ace.

afterwards. It's a lot easier to get results. Of course, the big idea is always to protect your own machines first and to worry about Huns next.

The show hasn't started yet, but should almost any day. When it does come it will probably make up for lost time. We may get a chance to do a bit of ground strafing and bombing which, according to the consensus of opinion, is a lot of fun and rather exciting.

From the letters that came today people at home seem to be rather worried about the push, which after all is probably a very good thing. But out here nobody seems to worry much, no doubt because we don't read so many sensational newspaper stories and magazine articles. The Huns can undoubtedly get certain objectives by paying for them, as they will have to, but the time is approaching when they'll have to think about the price. When the coin is men it's a terrible thing to squander, but apparently that is the least of Wilhelm's troubles. Everyone seems to think the coming push will be the largest and the last—westward.

When you get these you'll be home having a perfectly glorious time riding about the country and singing at Red Cross benefits.

Considerable love from
Bo

. . . .

[Beauvois]
32 Squadron, France
May 14, 1918

Miss Isabelle G. Young
434 West Sixth St.
Albany
Oregon
U.S.A.

Dearest Isabelle

Does today mean anything in your young life aside from merely being the fourteenth day of May, A.D. 1918? It's a large one to me. Just one year ago this evening I looked out of a drawing room window of the Lark headed south and saw Miss Isabelle G. Young. And I haven't seen her since. At least not materially.

But, Isabelle, there haven't been very many days in all that three hundred and sixty-five that I've not thot of you a little, and there have been a great many when I've thot of you a great deal. . . .

No patrols today. Headquarters must have lost our phone number, for it was perfect all afternoon and other squadrons were working. We played ball most of the time using an indoor baseball and a cricket bat. It's surely a treat to watch some of these Englishmen struggling with the intricacies of baseball, but they get on very well. It's really quite a good idea, for nobody takes half as much exercise as they should around here.

About five I indulged in a bit of mild aviation. Thot it would be a good thing to get used to altitude, so climbed up to 15,000 and played around for a while. The country for miles around was visible, but it was darned cold and lonesome. I sang at the top of my voice and whistled and talked to the machine but was glad to get down again. As many of our shows are done around 18,000 feet, one has to become accustomed to it, altho the only noticeable difference is a slight shortness of breath. Also if you don't swallow repeatedly when coming down, the pressure on your eardrums will produce a bad headache. Personally I busily work a bar of Spearmint as it facilitates the swallowing.

Enough. The anniversary has passed. I'm hoping and praying that when the next one rolls around, we may celebrate it together.

Good night, lover.

Bo

. . . .

[Beauvois]
32 Squadron, France
May 16, 1918

Dearest Isabelle

To begin with it is hotter than the well known hinges and has been for a couple of days. However two or three hot days seem to be the limit out here. It's beginning to cloud up a bit. We have located a swimming hole only a few miles away and are going over in a tender this afternoon, if we don't have to stand by for a show—which we probably will have to do.

Went on my first patrol yesterday—an offensive one about five miles over the line. Unfortunately the Huns aren't very easily offended, especially in the afternoon, and while there were a few about when we first went out it didn't take them long to head east. We roved around for an hour or so starting at 15,000 and working down to 6000, before we crossed and came home. Archie was fairly active but not very accurate. They shove up a lot of the stuff around you, and one dassn't fly straight for long. You will make a turn and see four or five bursts where you should have been. It explodes in little black clouds but can't be heard unless it's very close.

The area we patrolled is a shambles, no trees, great masses of shell torn ground, and demolished trench lines. In the sparsely shelled places the ground seems covered by great pockmarks—at least that's what it looks like from the air. The towns are only masses of ruins.

As it was a lovely day, the country for miles was visible. I managed to sort of get the lay of the land. There are few distinctive landmarks over here, and the roads are the only dependable guides, all of the main ones being white and straight and usually bordered by trees on both sides.

It wasn't a very exciting patrol, but quite good enough for the first one. I'll no doubt be fed up on excitement before long. . . .

Much love, dear, from
Bo

. . . .

[Beauvois]
32 Squadron, France
May 17, 1918

My Dear

Not much of a letter this evening for it's too hot even to think of writing. We were pulled out at three thirty this morning for a patrol that left the ground at four thirty. . . .

The chief feature of the last couple of days was an offensive patrol yesterday evening. As usual we had to go a long way over and even then didn't see much—at least I didn't—one never does at first. All at once our leader turned into the sun and started climbing. Like a good little boy I followed and still saw nothing. Suddenly at about 14,000 he turned east and dived, the rest of us close behind. And then I got an eyefull. Below us was one of our poor old photography busses, the pilot throwing it about every way he knew. Coming down on its tail were six Hun scouts. Six more sat up above. Our flight went down like a load of bricks on the first six—nasty looking little machines with big black crosses and painted in gaudy colors. The other six came down on us, and the rest of our bunch on them—sort of in layers. The first thing that came to my mind was having been told that when a Hun was on one of your machines' tails, open fire no matter what the range. You probably won't shoot him down, but he'll quickly lose interest in his target when a few bullets whizz by. So down I went motor full on, got a squirt at one little purple and white devil, and let him have both guns. In the meantime Hun tracers streaked by leaving thin smoke trails and darn near scaring me to death. I pulled out, did a climbing turn, half rolled and dived again. Just as a grand dog fight was about to start, one of the guns jammed and I pulled away. It was all over in less than a minute, for all of the Huns dived east and disappeared. At least all but two did. They won't dive again. All of our machines came home and the photography bus got away which was very lucky for him.

Everything happens so quickly in a scrap that one hasn't time to think, scarcely time to act. I surely was scared blue.

This morning two of our boys were missing. God knows where they are. Not a trace of them anywhere—nobody saw them leave the formation and they are simply gone. This afternoon I went over for a new machine, and it's mine alone—a wonder that's never been flown before. It sort of puffs you up to have them hand you a machine worth a good many thousand dollars and say, "It's yours. Fix it up to suit yourself." Tomorrow we re-rig it and fix up all the little details and paint it so it can't be seen.

If it isn't one thing from you it's another and today the helmet came. Tomorrow it starts working. It's really too hot to wear a fur helmet. This one should do the work nicely, furthermore it fits. And, Izzy dear, it isn't that which matters particularly, but that you made it for me yourself and maybe thot of me once in a while—not every stitch but say every time you dropped one.

Good night lover.
Bo

. . . .

[Beauvois]
32 Squadron, France
May 19, 1918

Dear Lover

This Sunday evening you're probably thinking what a grand moon it is, and I'll grant you, were I anywhere else I'd ask for no more perfect night—BUT—for the last two hours Hun night bombers have been buzzing overhead shooting up the countryside in general. Our Archie has been sending up a barrage and searchlights and flaming onions. Our own bombers have been going back and forth on jobs, and the noise is very annoying.

Horribly hot again all day. I've been working since this morning on the new bus. Took it on the range; sighted the guns; timed the gun gear; and was just about to take it up for an engine test when we discovered a defect in the petrol tank and had to change that.

Then when it looked as if we were thru for the day along came orders for a six o'clock patrol. We got back about eight thirty. Went quite high—to 19,000—and saw a few Huns, but couldn't get near them. However Hun archie was very good and kept us on the jump trying to avoid it. Some of it came uncomfortably close and rolled us about a bit, but nothing very effective.

Tomorrow you shall hear the strange tale of a chap we lost yesterday and what happened to him.

Good night, Izzy dear,

Bo

. . . .

[Beauvois]
32 Squadron, France
May 21, 1918

My Dear

... To put it candidly it's as hot as hell, providing that well known summer resort is all it's supposed to be. Everyone wanders about looking for a cool spot, and every cool spot harbors some eight million pestiferous little insects who fly into your mouth if you happen to sleep with it open, and who crawl and buzz and annoy you in a hundred ways. I was up just before lunch fooling around, no coat, no helmet, no collar, sleeves rolled up, and it was hotter in the air than on the ground. Of course it gets a bit chilly up along the ceiling, but around 2000 or 3000 it's unbearable.

At present we're standing by, which means there may or may not be a show at any time between now and seven. There was an early patrol this morning. Dragged us out at four thirty. I had a terrible old machine, mine not being quite ready, and couldn't keep up with the rest of the bunch, so fired a green light and went back home. Nothing happened over the line anyway.

Last night the unpopular moon was working overtime, and so were the bombers. Not so many Huns but any number of our own machines. Being orderly officer again I read letters to Polly, Dolly, and Molly about Uncle Bill's nettle rash, Cousin Agatha's asthma, the weather, and the new swimming pool. It seems the men go to a nearby town to cleanse their person about once a week, and they all wrote at least a page about it.

Concerning the strange story of the missing man there isn't much that can be said within the censor's limits. He was on a patrol the other morning and suddenly went down. Nobody knows just why, but he was out of control. At the time we were about ten miles over the line. Being a gentleman of some importance close inquiries were made from infantry and artillery. Finally someone was found who had seen the whole affair. He had apparently gained control of his machine and headed for the line, his engine dead. The poor devil fell short in no-man's-land by a matter of a few feet, the Huns firing everything they could get their hands on at him. When his machine crashed in a shell hole, he jumped out and dived into an old trench. Apparently it had been a communication trench. One end of it led to the German line, the other to ours. Of course this chap turned the wrong way and ran smack into the Hun outposts. Just a few more yards in his glide or a turn to the left in the trench and he wouldn't be a German prisoner now.

As a consequence I've a new room mate [Lieutenant E. M. Jarvis], a rather nice sort of a boy who unfortunately looks as if he'd lost the last friend. Honestly, Izzy, when I wake up in the morning and look over at him I can hardly keep from weeping. A more melancholy looking person can't exist.

That's all for now, lover. I think of you lots and lots and wish we might be together.

Considerable love, Izzy dear.

Bo

. . . .

[Beauvois]
32 Squadron, France
May 23, 1918

Dear Child

... Some of the boys in the squadron know the nurses in a Canadian hospital several miles from here and go to their dances given every week or so. I was going to one this week, but Monday night the gentle Hun came over, scattered pills all over the place, killed several of the nurses and some of the patients. Things like that make your blood boil, Isabelle. It's a fifty-fifty affair fighting them in the air or anywhere else but when they bomb hospitals—well, it simply shows what kind of a game we're up against. There has been plenty of night bombing lately, but apparently the Huns had a bad show over London Monday night, six machines down, and our people drop ten pills to their one. They never come over in the daytime. Our bombers do more damage than they do at night.

We did an uneventful escort this morning bringing some bombers back. Strangely enough they were right on time and came back where they were supposed to, so we were thru early. However, we came back and went low over the line looking for Hun two seaters of which there were none in sight. Coming home we ran into a bad storm which came in from the west and had to go very low to see our way. It's great fun going down almost to the treetops and then "hedge hopping" home everyone wide open and trying to get there first. When every one gets there at about the same time, they fly around waiting for a chance to land.

Tuesday afternoon [May 21] three of us got into a nasty scrape and were very lucky to get out of it. My flight commander is a chap named Simpson, a Captain and as fine a fellow as one could wish to know. Furthermore he has a reliable head and plenty of experience. We were doing an offensive patrol about twenty-five miles over the line at 18,000. Below us was another patrol. For almost two hours we tootled up and down looking for trouble, but no Huns, no Archie, no excitement of any kind. Finally three of us went down to look over a suspicious two seater. When we got back up the rest of the bunch had gone home. We were about to leave, when along came nine gaudy Hun scouts beneath the bunch that was under us. We flew large circles and watched. Being pitifully green I had no idea what was happening. The bunch below us fell for the trick and down they went on the Huns. I expected we'd go down and join the scrap for they were at it hot and heavy. Suddenly Simpson turned west, and we followed. Wow!! What a jolt. Slightly below us and so close that every little detail was visible were some dozen Huns all coming out of the sun. We were so far over they had been able to get west of us. These Huns were painted even worse than usual, every color and every combination and design. They all were covered with enormous black crosses. One machine I distinctly remember, a gorgeous orange creation, for while I was looking at him he turned and shot at me. Expecting Simpson to half roll and dive on at least one or two of these birds I looked up, which should always be done before diving, and directly above us were at least twenty more Huns. Honestly, Isabelle, they had us cold. A turn and dive by any one of them would have put him right on our tails. I kept looking up and behind waiting for these people to come down so I could move around a little and not be in the same place too long. Simpson didn't hesitate a moment. He went straight thru about ten more Fritzies, turned into the sun, shoved his nose down and, with his motor full, on headed for the line. Where were we? Just as close to his tail as we could get without hitting him. About twenty Huns followed us to the line and wasted a lot of good ammunition, but they didn't have a show once we were under way.

The whole business was a trap, gaudy decoys below and more machines high up to come down on you when you go down on the decoys. Thank Heaven Simpson had sense enough to know what it was.

I didn't have time or brains enough to be scared, but the other two fellows who realized were pretty well shaken up over it when we came home. Believe me, we were just pure plain lucky, or maybe there was some deeper cause.

However we're back so why worry.

Good night dear lady.

Bo

. . . .

[Beauvois]
32 Squadron, France
May 24, 1918

Dearest

It was as clear as a bell last night. We were down for an early show this morning, but somehow and from somewhere a beautiful rainstorm blew in. It has been pouring ever since. Nobody minds particularly, for when it rains we sleep late and may be just as lazy as we please. Between you and me I'm getting too useless for words. Every time I think of getting up regularly every morning and working industriously all day for the well known spondulicks [dollars] it makes me tired and sleepy. But at that it will be a great relief to be doing something useful again. This war is the most useless affair that has been staged for a long time.

After sleeping most of the morning Simpson came around suggesting we drag a few of the other sleepers out, so we went around from room to room using cold water and eau de cologne and talcum powder to good effect bringing the curses of the multitude upon us. At present about a dozen fellows are sitting here in front of a fire arguing and discussing every subject that happens to pop into their heads. For an hour I've been trying to write, but you know how them little things is, when everyone is jabbering and you are trying to concentrate.

Bushels of love,

Bo

. . . .

[Beauvois]
32 Squadron, France
May 26, 1918

Dear Lover

Another letter yesterday—something from you almost every day. You know everyone out here lives from mail to mail and from leave to leave.

They wait for the mail in the morning. If they don't happen to draw, they simply settle down and wait for tomorrow's mail. And as for leave, there's a chap named Green who is scheduled for his two weeks in a few days. All he does is pray for dud weather and no patrols. "Wouldn't it be bloody awful luck to get done-in a day or so before your leave started," he will say. "Afterwards? Oh, I don't care what happens afterwards." It's just like the last few days of a semester, you can't wait for them to slip by. However, I've nothing to worry about for it will be at least two months before my time rolls around.

As for the weather, Oregon mists most of the time and no shows. We may get one in a few minutes, and again we may not. We will be sitting at tea in the afternoon when the phone will ring. Deep silence. Oh's and ah's and "where will it be this afternoon," are murmured and everyone crosses their fingers and prays while the O.C. carries on a one-sided conversation with Wing Headquarters. You know how it is listening to one end of a telephone conversation. Sometimes it's a show and everyone groans. Others it's orders to wash out for the day whereupon, sighs of relief prevail. It's a bit late at present for orders to come thru, but one never can tell.

* * * * *

These five little asterisks signify a lapse of several hours. There was no patrol but we had to practice formation with two new fellows who are in the flight. They seem to be ignorant of many things. One of them is fair but the other plows thru the formation, overshoots the leader, and then turns and drops miles behind. Incidentally, he had all of us scared to death. There is nothing worse than to have some fellow you have no confidence in cutting fancy capers about ten feet from your tail. When you know how a fellow flies and what he is going to do it's easy. For instance the chap who flies across from me knows just what I'm going to do when we make a turn. I know what he's going to do, and we both do it. Our system is to cross over or change positions, the outside man always going under as he has more distance to make up. All of which is probably as clear as mud and as interesting as a lecture on Egyptian hieroglyphic. . . .

Izzy dear, it seems sort of egotistical, to be telling you what happens to me day by day, yea, minute by minute, but that's about all there is to write about out here.

Having heard my little speech of apology let us proceed with the story.

Friday night several of us went over to a Tank Corps station to see a show given by one of the several travelling companies, the gaieties these were called. They are really a very clever bunch, all men but with a couple of Dick Morgans among them. The show was worth going miles to see. Plenty of good music and a lot of clever comedy. . . .

Last night's movies were Douglas Fairbanks in something or other and two comedies taken in Hollywood. Every scene was familiar. There were a great many French pilots there also, and they enjoyed it all hugely. One of them apparently could read English. He would shout out all the headings. Our pictures are usually quite good. The squadron runs them four nights a week and makes enough money to pay for films which are furnished by the expeditionary force canteens.

This afternoon a ball game with one of the other squadrons. You see, lover, we have plenty of amusement in spite of the war.

Speaking of the war—the darn thing's overdue again. Apparently the weather has been holding Fritz up, but there's no doubt he'll stage another push and one that will equal the March show. When it comes we'll probably get it hot and heavy, for we have to get control of the upper air so our ground strafing machines can work. . . .

Lots and lots of love from
Bo

. . . .

[Beauvois]
32 Squadron, France
May 28, 1918

Dear Child

Seein' as how I've just put away a couple of pints of tea and have one hour before we leave the ground for an evening patrol, what is there today that might interest you? We went on an escort this morning and found the Hun archie active and accurate. There was a ceiling of clouds at about 12,000. No doubt we were nicely silhouetted against it for the "wonks" were frequent and uncomfortably close. We had to keep on the move most of the time. After our bombers crossed, Simpson and I went down looking for two seaters. At about 3000 Fritz sent up some archie that made my hair stand on end. We dived down to just a few hundred feet and tootled along the line. Finally we sighted a Hun doing artillery observation just about at the line and only a little over 700 feet up. We went after him, but he knew just how close to let us come and then away he'd go. Three times we chased him away, and three times he came back. The last time we got a few shots at him, but went quite a way over after him. Coming back we got well peppered by machine guns from the ground. It's great fun getting down low where you can look right down into the trenches and see what's going on.

Last night a nearby town was bombed and shelled at the same time, which was very annoying and not at all conducive to slumber.

Must go to work, lover. More very shortly and just plain old fashioned love from,
Bo

. . . .

[Beauvois]
32 Squadron, France
May 30, 1918

Izzy Dear

About five minutes ago it suddenly dawned upon me that today is Memorial Day, also that it's your birthday if the old memory fails me not, but I wouldn't swear to it. . . .

For the birthday all the good wishes and love in the world and selfish wish that I'll be nearer to you than I am now when the next one rolls around. As for Memorial Day, it means nothing in my young life. Were I in those dear United States it would signify a day of surcease from toil, a morning of luxurious slumber, and a double header ball game in the afternoon to say nothing of all the flags and brass bands and G.A.R. [Grand Army of the Republic] parades. But here—

At three thirty we were dragged out, then a few moments of silent cursing and stumbling into clothes, an egg and a cup of tea at the mess, and leave the ground at four thirty. But just about five minutes before we were to leave a magic fog sprang up from nowhere. We all went back to bed. However just at lunch time we were sent out and finally had lunch about three. Saw plenty of Huns but they were quite foxy and wouldn't let us near them except once, and then only a few shots were fired. Another show at five thirty and we were thru for the day at eight. Sure as fate will get another early patrol in the morning. You see, dear, the war is about to start again, has started in some places, and they are taking no chances. Anyway, five hours flying is plenty for one day. On top of that my poor old physiognomy is all chapped, and my lips are split from the wind and the cold up high and hot sunshine on the ground.

Yesterday we did nothing, as it was misty all day, but the afternoon before we had a bit of a fight and may have brought down two Huns. It was too misty to see what happened but two of 'em had so much lead pumped into them they should have been dead if they weren't.

And of course last night a lot of fool Fritzies came over laying eggs promiscuously over the countryside for hours thereby interfering with the wooing of Morpheus.

However it's a nice war and I like it.

Good night, lover,

Bo

7. MONTDIDIER AND NOYON *June 1918*

On May 27, the Germans attacked with thirty divisions, driving a twenty-five-mile-wide salient twelve miles into the Allied line. Green doughboys of the American Third Division were rushed to help form a defensive line on the south bank of the Marne near Château-Thierry. Paris was threatened.

On the first Sunday in June, in the middle of the night, Ninth Brigade headquarters ordered the Thirty-second Squadron to move seventy air miles south to an airfield in the French sector at Fouquerolles, which was near the town of Beauvais, well behind the front lines and north of Paris. On June 9, eleven German divisions renewed their offensive on a front that was less than sixty miles northeast of Paris. The French counterattacked on June 11. The Thirty-second Squadron concentrated on flying ground-strafing missions and escorting observation planes. Two days later, the German high command ended the offensive.

With relative quiet restored in the French sector, the Ninth Brigade was ordered north once again in anticipation of an attack on the British front. On June 21, the Thirty-second Squadron was pitching its tents at the Ruisseauville airfield near Fruges, only a few miles from its old airfield at Beauvois, where it returned to the mission of escorting day bombers.

. . . .

[Beauvois]
32 Squadron, France
June 2, 1918

Dearest Izzy

We have been ordered to be ready to move and have been packed for two days, but nothing seems to happen. Like Mohammet, if the war won't come to us, we shall go to the war. Apparently that's what is going to happen. If we move where we expect to we'll get a lot of hard work.

However, move or no move two shows a day has been the schedule, but they haven't been very exciting. Almost every patrol we see a few Huns and go down on one or two of them. Immediately they dive east in a terrible hurry. A couple of mornings ago there was a large formation of Triplanes in the sun to the east and above us. They tried all the tricks in the bag to get us to come over and play with them, but nobody bit. First they made believe there was a big scrap going, on all of them buzzing around and diving and spinning. After they tired of that, two came down below and tried to draw us east after them. But nobody shows any interest when the Huns have the advantage of numbers, height, the sun, and being miles on their side of the line.

However, all of that is trifling compared with what happened yesterday. Two letters from Albany and both of them perfectly splendid, dear [answers to the letter to her parents he mailed on April 5]. In fact I should be answering them now instead of writing to you.

Both letters were so awfully decent, more than that considering the unusual methods employed by Rogers. It surely takes a load off the old mind, and I'm afraid I'm going to like your folks an awful lot.

This evening I noticed on the calender that today is Sunday. Otherwise I'd never have noticed it for all days are alike over here, and one soon loses all track of time, except when we have to get up at three A.M. Most of the fellows here in the squadron advocate a union war, eight until five with an hour for lunch, half holiday Saturday, and Sunday off. However Sunday seems to be busier than other days.

This evening after supper [censored] and I took a long walk up to the next village and it's an evening to write home about. Everything seemed very still and peaceful and green, the trees just coming out nicely, the fields all planted and the grain beginning to grow, a perfectly well behaved first Sunday in June. But from the east comes always the dull rumble of the guns. Up there along the line there are no green fields and budding trees. They sow the fields with iron and reap—I wonder what the harvest is—probably sorrow for the most part.

It's a lovely war. . . .

It's late and I'm sleepy—you'd be too, if you'd done five hours up above and knew you were going to do five hours tomorrow and the next day and the day after that, etc. etc. ad infinitum.

And best to your mother and father and for Isabelle—ultra best.

Bo

. . . .

[Fouquerolles]
32 Squadron, France
June 6, 1918

Dear Child

Sunday night, the day of the last letter, I had just snuffed out the candle and was thinking of you, thinking myself to sleep, when heavy steps on the stairs and the door was rudely opened. 'Twas the C.O. "Rogers," says he. "Yes sir," says I. "Everything to be packed in half an hour. Only keep out what things you can carry in a haversack in your machine." And he departed. So I crawled out, dressed, packed all the chattels, and saw them on a truck which departed shortly thereafter. There surely was a lot of moving around and excitement. Everything had to be packed on the transport. About three a long train of lorries and tenders moved away. We went up to the aerodrome, crawled into flying suits, and climbed up on top of the hangers for a few hours sleep. In the morning we flew down here covering in about an hour a distance that took the transport fifteen. Of course there were a lot of things that turned up at the last moment. I came down with two large hams and a pair of boots tied to my bomb rack. There was nothing here but a deserted aerodrome when we arrived. We went into a nearby village and found a little estaminet where madame was delighted to see some strange troops and served much horrible wine and lemonade at a nominal charge. Later the O.C. took several of us into a town of some size about ten kilos away, where we had dinner and looked around at the sights. There is a fine cathedral and several large hospitals, among them an American one.

About six the transport arrived much the worse for dust. We began to settle down and arrange for a little comfort. We have tents as there are no good billets nearby and tents are really preferable during the hot weather. At present everything is pretty well settled, rugs on the floor, serviceable if not artistic wash stands built from oil cans outside the door, and all the comforts of home.

This is a beautiful part of France, much nicer than where we were. The war seems to have affected it less. There are many pretty towns, plenty of trees, and the people are delightful and can't do enough for us. Everyday they bring large bunches of roses for our mess, and we have no end of fresh vegetables, eggs, milk, and butter.

There is plenty of war here, and we are closer to it so don't lack excitement. Last night the Huns came over and scattered pills all over the place. The town with the hospitals put up a very effective archie barrage

A tender clearing away a crashed "A" flight plane.

and flaming onions. The Huns pop away at the searchlights with their machine guns, regular Fourth of July affair.

We did two shows yesterday. In the morning we chased two seaters, but in the afternoon we got into a stiff scrap with several Albatross' which tumbled down from above. [About four lines were erased by the censor here.]

He [Lieutenant Alvin Callendar] went down for several thousand feet belching flames and smoke and it surely looked like curtains for him but he dove the fire out and came home with nothing worse than a pale face and rather shaky nerves. [Another half-page was erased by the censor here.]

It surely is disappointing to miss out on such a chance, but there are plenty of Huns about and there should be scraps enough for everyone.

Do you remember the sad faced roommate I told you about? I'm in with another chap down here, but this boy went west this morning coming back from patrol. Just over the aerodrome something happened, he stalled I think and spun into the ground from about 200 feet. It was a nasty crash, and he died at the hospital.

Aside from these few little items there ain't no news, dear Isabelle. But there is another show this afternoon probably, and one must have one's tea.

Love
Bo

. . . .

[Fouguerolles]
France, June 9, 1918

Izzy Dear

Just a very short note for there isn't time for any more. There's a gosh awful war going on not far away and work to be done.

Honestly, dear, I saw things this morning that are beyond all description—just a hell of smoke and dust and bursting shells and gas clouds and blasted and burning villages. I thot I'd seen real action before, a small strafe on some road or trench, but if this is war take me home.

The Huns started a push last night, and at six this morning we went out on a road strafe carrying much ammunition and a couple of pills. Going fifteen miles over the line at 2000 feet, getting machine gunned from the ground and archied all over the place is not my idea of a peaceful Sunday. I dropped my pills on a road and got one direct hit right in the center. After that we dived at anything that looked worth while and pumped lead into it. You can't imagine what a shambles the whole place is. One immense shell burst in a village and smeared houses, street, everything into one smoking jumble. The towns are wrecks, many of them in flames. The roads and railroads are simply plowed with steel and planted with high explosives.

On top of all that we were practically flying thru a barrage of shells the Hun tracers streaking by uncomfortably close. It can't be described. Finally after many darn good attempts archie got to me and nearly ripped one aileron away, a great ragged hole that almost cut the controls and shattered the aileron. As my guns were empty, I couldn't see any percentage in hanging around so came home. One of our chaps was wounded, a gunshot in the leg, and another one gone west.*

It's all so terrible that I can't realize it, dear. And to think that down on the ground in that terrible mess are men—swell war.

However one mustn't get the wind up, especially when we'll probably get a couple of more shows like it today. A nice shave and a large breakfast have made the world a little brighter, but I'm really not keen for the next job.

This is all for now. I'll guarantee the next letter won't be so depressing. Much, much love from
Bo

---

*The wounded man was Lieutenant Robert Graham. Captain Walter A. "Bing" Tyrell, fifteen-victory ace, was killed.

. . . .
[Fouquerolles]
On Active Service
France, June 11, 1918

Dear Child

Another one of these disconnected letters about nothing in particular. We have had one show this morning, are going upon a second one in less than an hour, and a third sometime after that. The war is going again this morning only this time we are doing the shoving, and we've been rushed to death. Yesterday we had a show at four A.M., came back and went to bed again and slept until noon. A steady rain made it look as if we were thru for the day but not so. About six we went out on a job, rain and all, and it surely was a wet sloppy affair.

This afternoon we are on ground straffing again and every one likes it about as well as they do smallpox.

* * *

Denoting lapse of time—some eight hours to be exact.

While I was rambling along orders came thru that there were Huns prowling around on our side of the line shooting up two-seaters. So away we went. Four of our bunch went out to strafe, and six more of us were to patrol at 5000 to protect our two-seaters and low flying machines. The war was getting on quite well when we arrived at the line. Our people were making quite fair progress and it was very interesting to watch the battle on the ground. As it was a perfect afternoon we could see everything. The roads were filled with transport and troops; the field artillery was working overtime; and there were several tanks crawling along across the fields and over trenches. Pretty soon seven Huns popped across the line and we promptly chased them back. Then we zig-zagged down the line with the sun at our backs, and climbed into it again.

Our second bit of trouble came when five green and yellow Albatross' tumbled down on us from above, shot, and then dived away east. We went after them, and I managed to get on one's tail and get some good bursts into him but darned if I could shoot him down. Guess I haven't got the knack. I thot this merchant was cold meat, but he never faltered.

A few moments later a French two-seater working for his artillery was hopped on by two Huns, and we chased them away. The last excitement came just before we were leaving the line for home. A Hun two-seater came across right beneath us, then apparently saw us for he turned and dived. Two of our patrol tried to get under his tail, which is the proper way to attack a two-seater but I, being up above, went straight down on him. By golly, I was right on top of him banging away with both guns, and so close that sights weren't necessary—and he got away. Something seems to be wrong. Incidentally the observer put a few small holes thru my planes. So much for the war. . . .

Some of the lads have been into town for supper this evening and apparently imbibed too freely of grape juice which is plentiful and cheap in these parts. They carry on too hilariously and hold their liquor badly.

No more now, lover. Us laboring men must have our slumber.

I love you, Izzy.

All yours,

Bo

. . . .

[Fouquerolles]
On Active Service
France, June 13, 1918

Dear Lover

Here it is almost bedtime again, but there's always time for a letter to Isabelle. Besides it's not about midnight as you may suppose but about nine in the morning. We were coaxed out at three A.M., dressed in the dark, stumbled over to the mess for an egg and some coffee, somnambulated up to the aerodrome, waited for a ground mist to clear, and left the ground on an offensive patrol at five. When you start out on a show that early you're generally about half awake but two hours of fresh air and a lively scrap knock all the slumber out of you and when you get back you're unable to get to sleep again. Ain't it awful?

We had a lovely affair this morning and got one Hun [Albatross] in flames—not me, but another chap [Captain Claydon]. I managed to shoot one off of this fellow's tail and chased another down to 2000 feet which I later regretted as climbing back toward the line the Huns on the ground put up a barrage of nice incendiary and explosive bullets. It wouldn't be so bad if you couldn't see them but they look like streams of fire and put the wind up your old friend Rogers.

Last night this same fellow got another Hun—also in flames. They are plentiful around here and would be very likeable if they'd lay off of this unsociable habit of shooting at allied machines.

We've started working on a new system and it looks pretty good so far. Also we are going to wait for point blank range before opening fire unless it's a case of shooting Huns off someone's tail. If you open up at long range the Hun immediately spins or half rolls or gets out of range somehow. But if you wait until you're right on top of him before letting him have it, he hasn't time to get away.

Probably all this blood and thunder war talk gets tiresome but in these busy days it's our chief supply of news.

Considerable love.

Bo

. . . .

[Fouquerolles]
32 Squadron, France
June 14, 1918

Hello

Something apparently is wrong. Only one patrol yesterday and none as yet today. However it's too good to last, and there surely will be one before night even if the clouds are low and the visibility very poor.

The morning was spent in sweet slumber for the most part—and then getting my Lewis gun in shape. It had a bad habit of stopping just when it was most needed. The stop pawl was allowing the magazine to override causing double feeds, and one of the extractors was chipped. Figure that out!!

Then I'm having a new carburetor put on my kite and had to stand around just as if I knew as much about it as the mechanics.

Barring accidents you shall have a picture frame before very long. Yesterday morning diving on a Hun I shot my propeller thru. While they can usually be plugged up this one couldn't and should furnish any amount of souvenirs. There are any number of trinkets that can be made from a propeller, but few are small enough to get across the large ocean. One of the chief stunts is to take the hub, and after finishing it properly, set a clock in the center. We have a couple of men in the carpenter shop who are cabinet makers and who can make any amount of things.

Oh yes! Our chief form of amusement these days is ping pong. We had a large table made for the mess, and there was already a set here so whenever anyone craves excitement they engage in a strenuous game of ping pong. We're easily amused. . . .

All yours, lover.

Bo

. . . .

[Fouquerolles]
32 Squadron, France
June 16, 1918

My dear

Life in the army is apparently just one move after another. Two weeks ago today we moved down here. Now we are getting ready to move again, where I don't know, but doubtless where the war is to become more strenuous. It's sort of quiet around these parts now and not exciting enough for us.

We did only one show yesterday, none the day before, and one the day before that so you can see for yourself that we're not earning our pay. This morning we took a lot of bombers across. They were high, but we were higher, and a colder show I've never been on nor hope to be again. It didn't take long for the hands and feet to become absolutely numb, so it wasn't bad, but coming down they began to thaw out. They ached and itched and tingled and didn't get back to normal for hours. It was rather a nice morning in spite of the cold, very clear and a mass of fluffy white clouds below us. Nothing happened. . . .

I'm a regular old uncle these days, for a letter from mother yesterday stated that my sister [Adela Rogers St. Johns] has presented the family with a girl. She's a large hefty child with abnormal lungs and so cute, you know the old stuff, looks like her mother and is smart as a whip, two days old, laughs and gurgles and has an unusual head of hair and an enormous appetite. Personally I was pulling for a boy, but it can't be helped, and girls aren't exactly undesirable, at least some girls.

Unless something unusual happens again you shall have a picture frame before long. I hacked one out yesterday and am having it polished. It's not too good as the propeller was too even grained, but maybe a better one will show up. Don't expect too much of it for they are only curiosities at the best and look better over the kitchen sink than in the front parlor with all the Chippendale.

As for knitted socks I've worn nothing else for months and they haven't an equal for warmth, comfort, yea beauty. But look matters straight in the face, dear, and realize while my pedal extremities are not abnormally large neither are they excessively dainty and while sweaters will stretch socks invariably shrink, especially when one of these French farmer ladies washes them. Where the French laundries got their reputation is a mystery to me, for the proletariat can do more damage to your clothes than the old Red Star of San Jose ever dreamed of. Furthermore unless you have an exclusive laundress, yours and ten others' laundry will get beautifully mixed up.

When it comes back we have a sorting bee and try to get back what is rightfully ours. Somehow or other I always come out short a few articles. . . .

Always,
Bo

. . . .

[Fouquerolles]
32 Squadron, France
June 18, 1918

Dear Child

All of which reminds me that I've just moved over to a regular room with real floors and a roof and everything. One of the chaps I've been running around with [Wilfred Green] has been made a captain and flight commander. Flight commanders have preference for rooms in the one billet in the town. So he moved in and two of us came along. It's a very nice room and after all better than a tent.

We haven't moved and from the latest reports aren't going to for the present. Everything was packed and ready to go but nothing doing. I'm perfectly well satisfied with this place and don't crave hopping about from place to place. . . .

Things are back to normal again as far as work is concerned. Two shows today—one fairly exciting—and one yesterday. The last was an escort and while a bit tiresome was one of the best conducted affairs we've staged in some time. Thirty bombers went over on a little raid, and two scout squadrons escorted them. The bombers were below at about 12,000, one bunch of scouts slightly above them, and we sat on top at about 16,000. The bombers came across the line in three groups at about ten minute intervals. We would meet the first bunch at the line, take them over while they laid their eggs, bring them back to the line, and take the next bunch over. Everything went off without a hitch and woe to any Huns who might have come snooping around. The sky was full of our machines just like a big flock of sparrows. However the poor old bombers tootle right along, but have an enormous amount of archie shoved up at them, just get plastered with the stuff. They don't seem to mind it at all and never try to get out of the way. It's surprising how close that stuff can come without doing any damage.

More love than usual from,
Bo

. . . .

[Ruisseauville]
32 Squadron, France
June 21, 1918

Dear Little Lover

Here we are up very near to the place where we used to be and everything in a turmoil. We moved after all, yesterday the transport, and this morning we flew up. We are only a few miles from our old aerodrome and have quite a nice place altho we are under canvas again. The aerodrome and sheds have been used before and were in pretty good condition, but there were no officers quarters, and all afternoon we've been laboring to make our tent fit to live in.

First the floor had to be leveled and some of the grass removed, then the pegs all reset, they never are where they should be, and then we hunted around for furniture and now have a large bomb case for a table and two gasoline cases nailed to a tree make excellent wash stands. Furthermore as it looked like rain we dug a drain around the place. About five minutes after it was finished the clouds opened. It's been raining ever since, and this the exact middle of summer.

One gets quite adept at housekeeping tricks out here. You soon get so that no matter where you land you can have a cozy place of some sort rigged up in no time.

The day before yesterday there were four letters from you, all long ones and all—you know. I'll answer 'em verbatim next time.

Yesterday—ah! Long story. All of our stuff had left, and about four we were washed out for the day. The place where we were was closer to Paris than any other place we're likely to be for a long time, quite a little journey but not too long. There was only one tender left and while Paris is out of bounds we asked the C.O. if we could have it, and he unofficially said we could. He's the most regular person in the world and never refuses you anything within reason. So eight of us piled in and hied ourselves to gay Paree. Swell looking bunch of boys too. Nobody had any decent clothes for everything had gone with the transport. I had a fairly clean coat, but filthy, greasy breeches and field boots and most of the others were no better off, in fact most of them wore golf stockings and low shoes.

The sergeant who drove owns a ninety horse power Mercedes when he's home, so we rolled into the large town before seven.

You've heard all about Paris and all I can say is that it's that only more so. We went to the Hotel Royal for dinner after having driven around for a look at some of the more obvious sights and after that to Maxims. They say Paris now is dead, but if I'd ever hit the town before the war I know who would have been dead—and it wouldn't have been Paris. We really had a wonderful dinner AND ice cream, real strawberry ice cream, the first for many and many a day. The Café Royal isn't exactly quiet but Maxim's is a knock-em-down and drag-em-out joint. And the women—#@&*@&$—some of them are wonderfully dressed and very attractive after a fashion. Some of them are artificially beautiful. All of them use far too much drug store complexion, and all of them smoke, and there are a multitude. It was all very interesting and entertaining now that I know you, dear, beauty means something different, and to me you are supremely beautiful, Izzy.

We left about twelve and found it is much harder to get out of Paris than into it. "All roads lead to Paris" but all roads out of Paris don't lead to where you may wish to go. We made a good many inquiries before we got to St Denis and missed the road a couple of times after that.

It was raining this morning, but about noon it cleared enough for us to get away. We had to fly very low all the way because of the clouds. One chap crashed taking off and the C.O. ran into a sunken road landing here, his first crash in three years out here.

Oh yes—about Paris. There are Americans all over the place. I looked in vain for a familiar face, but didn't see a soul I knew. I'd have gone to the University Union had there been time but you know how them little things is. One of the chaps with us, an American [Reuben Paskill] from the University of Chicago, was for a visit there too, but we yielded to the sights.

No more now, dear Isabelle. The candle is just about gone, and if I don't go out and loosen the ropes all the tent pegs will be pulled up and then what? After I tuck myself in I'll read your letters.

Good night lover
Bo

. . . .

[Ruisseauville]
32 Squadron, France
June 23, 1918

'Lo Sweetie

Sumpin' always happens on Sunday, doesn't it. This morning we did an escort for a flock of bombers in one of the neatest little gales we've had in these parts for many a day. We left the ground at six thirty and got most of our height over the airdrome. As the wind was straight out of the west it took us no time at all to get to our objective, where we met the bombers.

There weren't any Huns about, but coming home we had an awful time. It took us about four times as long as usual to cover the distance and we were losing height from 15,000 feet, too.

After lunch we played tennis on our new court—or did I tell you about the new court? It's a very well behaved affair considering the fact that it was fixed up in a few hours and it's in use most of the day.

Probably some of the squadron mail went down where we were. If so it should find its way back up here in a day or so.

The articles in the "Post" by Billy Bishop [Canadian ace and holder of the Victoria Cross, with seventy-two aerial victories] that you've been reading aren't bad but he has painted them up considerably and sort of clouded up his popularity out here by writing them. There's no doubt that he's one of the best pilots that ever climbed into a machine, but writing all about it and telling how *I* did this and *I* did that isn't considered too good form out here. But they are vivid and fairly accurate.

What are you going to do next fall, dear? Back to Stanford or some place else or stay at home or what? As for Rogers—no doubt he'll be right out here in France next fall and heaven knows how much longer but the war has got to end some day AND WHEN IT DOES—. . . .

Did I tell you that Bill Taylor is missing? I wrote him a note and it was returned with "missing" scribbled on it. I wrote to another chap in his squadron for particulars but haven't had an answer yet. Of the four fellows I went up to Canada with I'm the only one out here. Quite cheerful. However Tom Whitman is in England and ought to be along shortly. Bill and Pop are gone and a chap named Leaf couldn't stand the altitude—had bad ears.

Anyway, I love you, Izzy, and I'm coming back to you and don't you forget it.

All yours.

Bo

. . . .

[Ruisseauville]
32 Squadron, France
June 25, 1918

Dearest Child

And that's all you really are, Izzy. Didn't I have a birthday yesterday . . . ? . . .

The birthday was not celebrated. It just came and went and the only event was rain. In the afternoon three of us went into town, got a much needed hair cut, and bought a lot of stuff which we ate while walking back to camp.

This afternoon we had the first show in a couple of days and got one Hun. Somebody else [Captain Claydon] got the Hun, a two seater, and all I got was a lot of tracer shoved at me by the observer. The way we do it is two fellows play around above the Hun and shoot at him occasionally to draw the observer's fire while one chap goes down underneath and does the dirty work.

There was a very strong west wind blowing, and after the scrap we went up above the clouds and were blown far over Hunland. We started for the line in formation, but when the leader dived thru the clouds everyone lost him and we got badly split up. Then the archie started and the sky was simply full of it. We were bucking the wind and not making much speed, so the Hun simply plastered us with it. I hadn't the least idea where we were but headed northwest figuring the wind had carried us quite a way. For about ten minutes I tried dodging the bursts but gave up in disgust and flew straight, only changing my height regularly, climbing and then side slipping. Some of the stuff certainly came close, and it was always plentiful and regular. Finally I crossed the line and recognized where I was, over thirty miles south of where we crossed going out. A couple of other kites came along, and the three of us came home together. Everyone got back after looking around the country — one chap crashed in another aerodrome.

The weather sort of promises not to be too good for a while, so maybe we'll get a bit of a rest. I hope so for we have a lot of new pilots in the squadron, and it will give them a chance to work in. Simpson and I are the only old men in our flight and heaven knows I'm anything but old, only comparatively, only comparatively.

Show in the morning. Hour now is nearly twelve.

Good night, lover,

Bo

. . . .

[Ruisseauville]
32 Squadron, France
June 27, 1918

Dearest Isabelle

Yesterday a letter to you was in embryo, should and would have been written in the evening, BUT we had a show wished on us at the unreasonable hour of seven fifteen, didn't get home until nine thirty, and after dinner I had to censor the mail of which there was a vast quantity. In the midst of it came orders for a show at dawn so you see—c'est la guerre.

Last night's patrol was a great affair. We were to bring back some bombers and by the time we had escorted them across the line it was nine o'clock and getting dark. We headed for home in a hurry and there was one road we were able to follow until it ended at a town a few miles from the aerodrome. From there it's a sort of cross lot trip, and it was pretty hard to see where we were going. Then we started firing lights, both to keep the flight together and to attract attention on the aerodrome. They sent up a lot of lights in return and everyone got back safely. One chap crashed as it was quite dark, and I came too close to scuttling my kite for comfort, hit with an awful bump and bounced all over the place.

This morning's show wasn't as early as we expected owing to an early ground mist, but when we finally did get up it was wonderful, clear and warm, archie very poor, and no Huns about. Yesterday for the first time I carried a nice red and yellow streamer on my tail as Deputy leader of our flight.*

At present I've completed the daily exercise, a couple of sets of tennis, helped give a little white puppy his first bath, and am sprawled out on my tummy in the sun writing this. It's a glorious day, dear, and I can't help but think how nice it would be roaming the hills somewhere with you, or driving, or sitting in a hammock, or just being with you anywhere. And don't forget that these lovely days are synonymous with two patrols. This is one occupation where good weather is a curse and stormy days a blessing.

We really lead a very prosaic life out here. About the only things worth writing about are unimportant. But there's one thing that *is* important—at least to me. I love you.

'Bye, dear,
Bo

. . . .

---

*Since Rogers joined the squadron in early May, eight pilots had been killed or wounded or were missing. Two more had fallen sick. Rogers was named deputy flight leader and now carried streamers attached to his tail, so that new pilots could spot them after a scrap and follow him in the absence of the leader.

[Ruisseauville]
32 Squadron, France
June 29, 1918

My Dear

The socks are wonders, by far the nicest of the heterogeneous collection I've rounded up. They fit. I'm glad you have no foolish notions about size. . . .

We did only one patrol yesterday, an escort for bombers in a lot of huge summer clouds. Coming home we had a bit of a scrap. Four Huns came down from the northwest in the sun and about a thousand feet above us. Two of our bunch—there were seven altogether—were up top, and the Huns went after them. The two tried to draw the Huns down under us. We had turned and were climbing, but they wouldn't come even to our level.

When we finally did get up to them they dived away into a cloud. The Huns were all in Fokker Biplanes [D-7s], the newest thing they have in the scout line, and if they were putting up their best performance yesterday we have it over them in speed, climb, and maneuverability. However they are apparently handled by good pilots.

Just at present the bombers are taking off, which means we are due for an escort job in about an hour. The bombers always leave long before we do and climb and climb to get their height. Then we take off, get up to them in a few minutes, and take them across. It's the same coming home. We trail them over the line, and when they are well out of danger we pass them and are all washed up by the time they land. You see there are advantages to flying scouts.

Faintness overcomes me, dear Isabelle. Must be time for tea. Tomorrow is Sunday and there will surely be something to write about then.

Just plain yours,
Bo

8. SECOND BATTLE OF THE MARNE: CHÂTEAU-THIERRY *July 1918*

By the end of June, Bogart Rogers had logged nearly 100 hours of war flying. He was a cautious pilot, careful and deliberate. He stayed with the formation, fussed over the new pilots, and flew top rear, watching for trouble and protecting the bombers. Years after the war, he would write,

> Most fliers, I think, were fatalists. It was the only doctrine that would hold water. If you embraced it, as many did, it was a great source of consolation. You simply decided your destiny was predetermined and inevitable and ceased worrying about what might happen to you. When your time came it would come—there was nothing you could do to stop it.*

For the first dozen days in July, the squadron flew bomber escort at the northern end of the line where the British were expecting a German offensive aimed at Flanders. The day-bombing ended abruptly on July 13, when eight squadrons of the RAF Ninth Brigade were once again ordered south to the French front. The Germans were moving their assault units into position along the Marne near Château-Thierry.

The squadron flew to Touquin, as close to Paris as it would come. At midnight on Sunday, July 14, the thunder and rumble and flash of artillery awakened Parisians. The German attack had begun. More than 2000 guns flamed along a fifty-mile front. A counterattack by French and American forces on July 18 surprised the Germans by its ferocity and forced them to abandon their bridgehead across the Marne.

. . . .

---

*Bogart Rogers, "The Startling Truth About War Fliers," *Popular Aviation*, December 1930.

[Ruisseauville]
32 Squadron, France
July 1, 1918

Miss Isabelle G. Young
Box 615
Newport Oregon
U.S.A.

Dearest Izzy

For the last four days we've done the same show in the same place at the same time with the same bunch of bombers. This morning we had a climbing contest with six Huns, but they were above at the start and also at the finish. They went down on our bombers over the objective, but lost one machine for their pains. We never go all the way with the bombers, but patrol between them and the line. Under ordinary conditions they can take care of themselves, especially in a running fight as a good observer can get in a lot of accurate shooting with his rear gun, and when they keep in close formation all the observers are able to converge their fire on one or two Huns. A good two seater is a bad thing to tackle.

Yesterday afternoon we ran into four Huns, two triplanes and two Fokker Biplanes, and had a little tiff but nothing important happened.

In about half an hour we are going upon our own, a roving commission with nobody to worry about. Our instructions are to get Huns. I'm going with six fellows on top as deputy leader, and we'll probably spend most of our time about 20,000 where it's a bit cool and the old kite very sloppy on the controls. This morning we were around 18,000 and the old bus wobbled and slipped all over the place.

Here it is the first of July and really a July day for a change. All the Canadians are celebrating Dominion Day, and there probably will be a large binge tonight. But wait until Thursday—that's the old day to celebrate.

Must put on the sweater and the flying boots and away for a crash into the atmosphere.

Much, much love to Isabelle,
Bo

. . . .

[Ruisseauville]
32 Squadron, France
July 3, 1918

Dearest

... [I]t's preposterous to say that there ain't no such word as "popcorn" in Italian. Just look at the facts. I personally am acquainted with some thirty-two popcorn venders in the city of Los Angeles and its suburbs. Of these thirty-two merchants one is a Greek, one an Armenian, one a half breed Turk, and the other twenty-nine thorough bred Italians, the blood of Caesars flowing in their veins, the indelible marks of Rome upon their swarthy faces. Do you suppose for one single minute that these twenty-nine honest tradesmen are lacking the word "popcorn," the means of their livelihood, in their native vocabularies? Taint possible; taint common sense. You've no idea what a vital part popcorn plays in my young life, lover. . . .

Nothing doing today as it's very windy and cloudy. It may clear up before evening but I think not. Yesterday we did only one show. The bombers pushed us up to high heaven. There wasn't a Hun in sight, and I was completely disgusted. Jerry Flynn, he's my young tent mate and deputy leader of another flight, says he knows where there are a lot of Hun two-seaters, so we're going down together when we get a chance and see if we can scare up a couple. There have been practically no Huns here for days, but some fine morning they'll probably appear in swarms and raise the dickens. As Simpson says, "There's only one thing worse than seeing no Huns at all and that's running into about twice as many as you can handle."

Last night we had a lot of guests to dinner, much champagne and "a pleasant time was enjoyed by all." Tomorrow is the Fourth of July, and you can figure out for yourself what will happen. There are two attached Americans in the squadron, and a couple of others besides myself from the States, and we're throwing a party. We will celebrate, we shall.

It's not very nice to over celebrate, Izzy dear, and get slightly wonky, but it gets so terribly monotonous around here at times that anything is a relief. And there really are a lot worse things one can do. You understand, don't you, lover? . . .

Heigh ho! Lovely war, madam, lovely war. Wonder when the damn thing's going to end.

All yours,
Bo

. . . .

[Ruisseauville]
32 Squadron, France
July 6, 1918

Dear Lover

Altho we've stood by all day waiting for a show there have been none. Standing by is a useless sort of thing. You get orders to stand by from such and such a time, and all you can do is sit around and wait, can't go calling or run into town or do any thing that you'd like to do.

We had a great party on the Fourth, lots of guests included a couple of majors commanding the bombers we take care of, the wing doctor, the wing staff captain, three or four flight commanders from the bombers, a major from the tank corps, and a couple of Americans from a corps base which isn't far from here. The mess was all decorated with flag and flowers and the menu — crab salad, soup, fish, chicken, cauliflower, asparagus, strawberries, nuts, and coffee — to say nothing of liquid refreshment. It was a great young party and lasted until about two A.M. A couple of the bombers had to go out and take pictures at three-thirty, and it looked very much as if we'd be on about six but Providence intervened and sent many heavy clouds over which enabled everyone to sleep until noon. Thank goodness the Fourth comes but once a year.

On the afternoon of the Fourth we saw many Huns and had a couple of good scraps. I was flying in the rear and above our flight and spied a couple of Hun two-seaters taking pictures about 2,000 feet above me. I went up after them, chased them east, but couldn't get close enough for any good shooting as they were too high. While I was gone the rest of the flight had a great dog fight with some Hun scouts and brought one down, maybe two. Everyone had some sort of trouble that day and it looks as if maybe things would liven up a bit around here.

That's all, Izzy dear — except much love from,
Bo

. . . .

[Ruisseauville]
July 8, 1918

Dear Child

Rather a busy day this — at least up to now and there's a possibility of another show this afternoon. This morning we took the bombers across quite early, and when they were coming back about a dozen Huns came up after them. There were other of our machines out at the time and a large and wild dog fight ensued. For the last few patrols I've been carrying one streamer and flying at the back of the top flight trying to keep a lot of new fellows in decent formation. When they get where they shouldn't be I promptly dive on them and rattle off a few shots past their tail as a gentle reminder. Anyway the top rear isn't a very offensive place to be, as you can't get down in time for a short scrap. When this melee started our bottom crowd tumbled down into it, and we stayed up top, the idea being that if any Huns go down on our people, we can drop down on the Huns. There were plenty of our machines below, all tearing about wildly and shooting like fury, so we sat up and looked on. Then Simpson saw a good chance and tumbled down on a Hun, and I was left with three new men to take care of. Then I spied three red triplanes sitting at about our level and east of the scrap, so around and around we went watching the tripes. They wouldn't come close to us or to the scrap, but woe to the straggler who might come their way. They wouldn't go down on anybody as long as we stayed up—so we stayed up.

Finally the Huns dived away east, and the battle was over. From the latest dope five Huns went down. I saw one triplane crash, and three of our machines are missing, only one from our squadron, however.

Then later in the morning I went into town and spent a merry hour with the dentist. The result is that I resemble an ad for Pebeco. This dentist is an American attached to corp headquarters, and a most painless person, filled a good sized cavity for me and not so much as one little pain.

Yesterday we did a show in heavy clouds, and nothing at all happened. But when the bombers were taking off, the thirteenth and last machine stalled, crashed nose in from about a hundred feet, and burst into flames. About three minutes later the two bombs he was carrying exploded and made an awful mess of things. Wrote off one of our machines and nearly ruined the aerodrome. It was an awful spectacle, the worst I've seen.

Simpson has just wandered in, says send his kindest regards, aye his love, and also state that Simpson and Rogers decisively defeated Flynn and Green in two fast sets of tennis yesterday.

The picture frame has gone at last and I hope you eventually get it. It's made from the tip of a propeller that I shot thru while diving too fast on a Hun and has seen quite a few hours over the line. The little whatnots on the back are parts of aeroplane, and the cloth under the back is like the piece I sent before from a Hun two seater.

That's all, Izzy dear, except that I think of you a lot and love you like the very dickens.

All yours,
Bo

. . . .   [Ruisseauville]
                                                     July 10, 1918

Dearest Izzy

Here's me once more but up in the orderly room this time and just finished with the censoring of many letters.

The weather has been wicked, rain, wind, and clouds. Yesterday we did nothing. Tonight about six thirty a sort of a rift in the clouds blew over, and the wing very wisely thot we should work. A grand storm was coming up from the west and everything looked very bad, but orders is orders and away we went. At the line it was clear up to 10,000 but being right under the clouds ole John Archie was at his best and just plastered us with big black shells. The storm kept drifting toward us like a great black wall and underneath this mass of clouds it was like night. We played around dodging archie and seeing nothing until the clouds were at the line, then we started home.

We had to go down to 1000 feet to get under the clouds, and when we finally did get below what we saw was enough to make up for the rest of the show. It was just like being in a great cave. All around us it was dark, the ground was hardly visible and the engine exhausts showed up like streams of fire. Ahead and to the west it was light, the sun shining and a great rainbow off to the north. Behind this mass of clouds was like a solid wall shutting out everything. Then we ran into a heavy rain and I just hunched down in my office, pulled in my head, and let the water stream from the windshield. Finally we came out into the open again and streaked for home. It was terribly bumpy all the way—just like a Ford, and we were tossed all over the shop. Unfortunately two new pilots got lost and crashed at other aerodromes.

The new pilots we've been getting lately are not to be described in polite society. They don't seem to have the remotest idea of how to fly and crash upon every occasion. It's most discouraging, especially as the silly devils seem to think it's great fun. The mechanics are the people who come out on the small end, as every crash means a lot of work for them. They're a pretty fair lot of fellows who know their jobs and do them, but these habitual crashers are about as popular as the measles with them. I've two nice boys on my old kite, and they take a lot of interest in it, keep it as clean as your Packard. The engine is probably a lot cleaner as that's what really counts. They are always doing some little thing to it.

Tomorrow

This is being written tomorrow afternoon in respect to the rest of the letter. We did an escort for our bombers this morning, left the ground at six, went to 18,000, and nearly froze for two hours so my left thumb hasn't thawed out yet. Nothing in particular happened; it was very clear, visibility good, and we could see the pills bursting all around the railroad junction which was the objective. All the way home it was bumpy as sin and after landing very nicely I taxied merrily toward the hangars, hit a bump and a gust of wind at the same time, and smashed an aileron, a small matter but I'm afraid I'll get into this breaking habit.

The C.O. is away on leave; Simpson is in charge of the squadron; and your old lover commands A flight. By golly, things happen out here, but of course it's only temporary. Jerry Flynn, my small tent mate, has just been made a captain and has a flight of his own. We lost a flight commander in the scrap I mentioned in the last letter. Jerry's only nineteen but he's been out here seven months, knows the game, and is a stout hearted little devil. He's going on leave in a day or so, and I go when he comes back. That two weeks in town is surely going to be hard to take.

There's a rumor that we may move again. It wouldn't surprise me in the least. We seem to perch around the country regularly. Every one is getting used to it altho nobody likes it. Simpson says he can handle a squadron that stays in one place and doesn't do much work, but if he has to conduct a move he's going to blight his young life with worry and responsibility. Maybe we won't really go AND maybe we will.

A couple of nights ago four of us went down to another squadron about thirty miles from here for dinner. We had a very fair time, saw several fellows we knew, and the ride was lovely, as we went in the C.O.'s car, a fancy seven passenger high-powered affair with fur robes and such....

Guess I'll stop and see how things are getting along in my flight—get that. I'm just as crazy about you as ever, Izzy dear.

Just yours,
Bo

. . . .

[Ruisseauville]
France, July 13, 1918

Dear Child

We're up to the old tricks again, the transport has just departed and we're hanging around waiting for orders to take the airships down. From all indications the war is about to start again and we're going to the lively spot. I shouldn't be at all surprised if we were handed the sweet and popular old job of trench strafing altho everyone is praying "agin it."

There's no denying that we see the country this way for this time we are going to an entirely new and different place but it's wearing on the nerves. Poor old Simpson has died seventeen natural deaths in the last twenty four hours trying to decide what is to be done about this and that and finally saw the transport under way with cheers. They have quite a journey which may or may not be pleasant as it's sort of rainish and may pour before they get where they're going. Anyway there won't be dust to worry about.

I've been pretty busy myself[,] the—ahem—cares and responsibilities of getting a flight under way resting heavily on the old shoulders. The real trouble is yet to come, for if I don't get lost leading these birds to the new roost some of them will, to say nothing of the possibilities of their crashing on a new aerodrome. I would that the airships had gone down on lorries.

There's not much else to talk about just now. A chap named Paskill—Chicago, U.S.A., U.S. Aviation Service, attached to RFC—and I are going into American corps headquarters and invite ourselves to meals. We know several fellows in there, and our mess left with the convoy this morning.

Don't be surprised if there aren't any letters for a week or so. A move plays the dickens with our mail, and this one will be worse than the others. The worst part of it is that there won't be any mail from you.

Much love from,
Bo

. . . .

[Touquin]
July 18, 1918

Dearest Izzy

Thank goodness I'm finally able to take my pen in hand once more. We've certainly been rushed during the last few days. Even now we are standing by for a job.

We were dragged out for an escort this morning at four and haven't gone up yet. By the time we got up here to the aerodrome it had become quite cloudy and everything was called off, or at least left hanging and nobody knows when we're going. However bright and early there was a letter from you, the first in a long time and probably one of the last from Stanford. It came just in the nick of time for the morale of the troops was very low. However I suppose it's best to begin at the beginning.

Sunday morning we cranked up the airships and started on our way intending to make the whole trip without a stop if possible. We bucked a

Lieutenant Reuben Lee Paskill of the United States Air Service, who trained with Rogers.

"A" flight before a patrol, Château-Thierry, July 1918.

stiff wind from the start and ran into a heavy rain storm after about an hour. Nobody could see the way. Since it was obvious we would have to land for petrol anyway, we perched on a convenient French drome.

It was pouring by the time we landed and after arranging for petrol and oil three of us went up to one of the messes for lunch. As it happened Sunday was July 14th [Bastille Day] and the Frenchmen were celebrating. The first thing they did was produce much liquor. We had to fly the kites and were very, very careful about the amount we consumed. The Frenchmen were nearly insulted. Then they started on a long apology for lunch. Lunch was merely a pick up meal designed to keep body and soul intact. Dinner!! Ah, dinner was *the* meal. But lunch, they were almost ashamed to ask us to stay. We were all figuring on a couple of dry crusts and a glass of water but they produced a six or seven course meal, everything and the trimmings to say nothing of the liquid refreshments. Then the weather cleared a bit, and we had to get under way. They're a great bunch, these French squadrons, and I nearly died of fright when a murderous looking mechanic with a red handkerchief around his head started fooling around my petrol tank smoking a large and fiery pipe.

Monday morning bright and early we were up to the old tricks again, ground strafed the Hun a bit and repeated again in the afternoon. Two shows Tuesday, one yesterday, and very likely two today. And when we aren't really flying we stand by for a show which is just as bad.

We are very comfortably billeted in a village near the aerodrome and have a large house for the mess. I've a large corner room over the estaminet on the main street but it's the noisiest corner I've ever run across. Things happen all night long.

It's been as hot as sin for the last several days, sultry oppressive heat and frequent thunder showers. Today, however, the air has cleared out considerably, and it's very pleasant.

Sunday night I went over to an American squadron which is nearby and ran into several fellows I knew in Texas, also a chap named Jones, a DKE from Stanford who was there in '12 and '13 and who knows about everyone I do.

Monday we had a chap wounded ground straffing, and yesterday we went into a nearby town to see him. I surely woke up to the fact that the Americans are in the war now and in it right up to the ears. At the American hospital there were fresh casualties from the push. A lot of them were pretty badly off. We don't see much of that sort of thing flying and it's always something of a shock to see nasty wounds and blood and wild-eyed unshaven men who have just been thru hell. Enough—must do a show. I'll finish later.

This is much later, Izzy dear, and a lot of things have happened. The morning show consisted of a protective patrol for artillery machines doing contact patrol. Our people staged a big push this morning. It will be ancient history by the time you get this, but it was a very successful show according to the latest reports. I led three other fellows up and down a small sector of line, quite low and the two seaters were working below us. There was a good deal of war going on, heavy shelling, tanks in action, and fires everywhere. It was all very interesting to watch, but when you're leading a bunch over a perfectly strange sector and have to keep watching where you are besides watching for Huns and keeping a formation together it's none too easy. There were some Huns out, but they didn't bother us.

This afternoon we did another protective patrol, more machines and at a very bad altitude as we were in a mass of clouds. I was on the tail of the lower bunch but one chap—one of the crashers I told you about—wouldn't stay where he belonged, flew wide and very foolishly all the way to the line. We crossed in a bad wind, one that was blowing toward Hunland, and dodged in and out among the clouds. We hadn't any more than crossed when two solitary bursts of archie appeared above and ahead. Right away I started looking for Huns and in about ten seconds twelve of them tumbled out over a big white cloud, and down they came. This one chap straggling out in the rear was cold turkey for them and to make matters worse he flew straight as a string, maybe he didn't even see them. But I'm pretty sure they got him.

One Fokker biplane sat right on his tail and pumped smoke tracers into him until he suddenly went up in a big zoom and then dived. I tried to get around to him but immediately three Huns jumped on me and wasted a lot of high explosives. It surely put the wind up me to have these Germans tumble down and when they're above and outnumber you the only thing to do is clear out.

Enclosed are Nenette and Rintintius. They are sort of mascot affairs and the rage in Paris just now—every one has a pair of them and carries them around on their wrists and such. You know the old fad stuff.

It's very late, lover, and it's quite possible we'll have another early performance.

Just one bushel of love,
Bo

. . . .

[Touquin]
France, July 21, 1918

Hello Lover

Here it is Sunday again and for some strange reason it looks as if it might be a perfectly well behaved and uneventful Sabbath. No show as yet—it's about two P.M.—coming up a bit dud and we *may* not get one, altho it's probable that we'll have to take some bombers across later. We are having a lot of chaps from the American squadron over to dinner, and it's just our luck to have to leave the ground about seven and spoil the party.

The allied push is still under way and the line continues to advance in the right direction. Of course it's not a terribly large action and doesn't call for any undue rejoicing, but it's the first respectable gain our people have made in a long time and it's immensely encouraging, to the people out here especially. The war's a long way from being over, and the Hun has probably got a couple of good offensives left in him but maybe this will be the turning point. I hope so. I suppose a lot of people at home think it's all over but the shouting, but they want to remember there's a lot of good weather left yet, and Fritz will take advantage of it. I doubt if they'll be able to show much next spring.

By golly, Izzy, it must be wonderful out west now. It's very nice here, too, but what use is good weather when it always means two shows a day and when the person you'd like to be enjoying it with is away over on the other side of the world. Mother sent some pictures of my niece—or it's our niece, isn't it—taken at the beach and they almost made me weep. Jones, the Stanford DKE I told you about, and I decided the other night that when the war was over we'd join the rush to the west and once we got there we wouldn't never join no more rushes no place, no time.

Had a letter from a boy from home named Allan Crary a day or so ago. Allan is in England, a lieutenant in the aviation corps, and we're planning a large reunion when I go on leave. Funny how people turn up out here.

Another Stanford fellow, Bill Gibson who I may have spoken of before, rolled in here a day or so ago and is in another squadron. We had quite a pow-wow, and he borrowed the Quad [Stanford yearbook] to peruse.

The leave is coming closer and closer. I suppose I'll have to start having engine trouble and such. Everyone seems to get sort of nervous and cold footed just before their leave is due. You see them coming back from shows with all kinds of things the matter. Fact is, my old kite is getting pretty wonky and is due for a new engine and an overhauling. It's done almost 120 hours with scarcely no repairs, and the average life of a motor is about fifty or sixty hours. Once they get running nicely you hate to take a chance with a new one for you never know what the new one will be like.

Funny that the censor removed all the fancy lining from the envelopes. Ordinarily most officers' mail goes thru without anything happening to it. As for incoming mail I don't believe more than four or five letters have been censored since I've been out here.

I've some new pilots to send up this afternoon and have to dash up to the aerodrome and tell them what to do and what not to do. You see, us flight commanders packs an awful lot of responsibility.

And don't forget about that—I'm in love with you, Izzy.

Bo

. . . .

[Touquin]
France, July 24, 1918

Dear Izzy

I was just in the act of being about to start a letter to you yesterday when we were washed out for the day and I was dragged off to Paris.

Lots of things happened in Paris. First, Rube Paskill and I went down to the American University Union, put our names on the book, and looked thru the registers. There are dozens and dozens of Stanford people out here apparently and it seemed that I knew almost all of them. And a lot more who are out here aren't registered. After that we met the rest of our gang—there were eight of us altogether at Maxim's and had dinner. The celebrated Lieut. [René] Fonck was there in all his medals and glory and surely is a popular idol around Paree. He's the leading French pilot at present with fifty-six Huns and just about as many decorations. After supper we went to the Casino, a sort of a combination theatre-café. After the show we were strolling out when who should I bump into but Bill Reagan, Stanford '19, and a Chi Psi. I knew Bill pretty well at school and was surely glad to see him. About a minute later along strolls George Zacharias, a DU who you probably know. Knowing that he and Johnny Jeffers are as thick as a couple of thieves, I asked for Jef and they produced him after a short hunt. Zach and Johnny are in one of the American squadrons which are right on the next aerodrome. We're going to start making party calls. Bill is tangled up in some sort of an ambulance affair and hangs out in Paris.

We rode home in a nice seven passenger Packard belonging to one of the American squadrons.

Yesterday was the first day we haven't done a show for a long time. It rained. The day before we took the bombers across and ended up in a gosh awful dogfight. Incidentally I got my first Hun. The bombers were attacked over their objective by a dozen Fokker Biplanes. We were well above and tumbled down on them like a load of bricks. There were two particularly gaudy ones that caught my eye and I went for one of them and got in a couple of good bursts at close range. He pulled up in a slow stall, turned over on his back, and went down obviously out of control. I couldn't watch him very long for everyone was dashing around like mad, but some other fellows saw him go down. The other red and white one with a bright red fuselage and a white tail and wings was a stout devil and put up a fine scrap until Green got him. We managed to get four in the whole show which isn't bad. Coming home archie got a direct hit on one of the bombers. It went down a mass of flames, not a pretty sight, especially as one of the chaps jumped out at about 12,000 ft.

This morning we went out on an offensive patrol, twelve machines in two flights, six above and six below. Simpson had the upper lot, and I was on the tail. In between we had four new fellows. Simpson had to come home with engine trouble so I took the lead. There were four Huns well below us. I fooled around and finally got right in the sun and had them between us and the line. Then I waggled the wings and tumbled down. One Hun was sort of apart from the rest so I went after him and got in a couple of bursts. He spun down, but finally pulled out. Then I looked back and not one of these poor fish had come down. I couldn't see any percentage in playing around with three Huns so pulled out and climbed back to the rest of the flight. It surely is disappointing to get everything fixed just so, and then have a lot of poor nuts fail to back you up.

Coming home we spotted a Boche two-seater working on the line and chased him east.

Must dash away, lover, and brace up the nervous system with a cup of tea. You know us English—.

The leave comes in three or four days—cheers.

"Curtain please, Mr. Stage Manager"—'cept, I love you, Izzy dear.

Bo

. . . .

[Touquin]
France, July 27, 1918

Dearest Izzy

Had a letter yesterday morning—the one telling about feeding the army. You know I'd be perfectly willing to be drafted and sent from California to Camp Lewis if they'd guarantee a stopover for dinner at Albany and Miss Isabelle Young to carry in the food. You see, war *has* its redeeming features.

Yesterday we had an early show, quite uneventfully escorted a lot of bombers while they put a Hun aerodrome out of business, and were washed out for the day. A lot of fellows went into Paris, but I took my kite and went over to visit Johnny Jeffers. Found a lot more people I knew in the same squadron—Joe Eastman, a Stanford S.A.E., and a couple of fellows who were in Texas. It started to rain like sin when I was there, so I had to come home in a nice thunder shower. Today it has been very cloudy. Altho six of us have been standing by for a two-seater strafing patrol, I guess everything is cold.

I'm in the old depth, Izzy. The Justicia which was sunk a couple of days ago must have had mail on it, either incoming or departing. If it was coming from the states, seven million horrors, I probably won't have another letter for weeks. If it was leaving it probably carried a few of my efforts with the pen, a small matter in itself, but I hate to have them end up on the bottom of the ocean.

Leave hangs fire. Two fellows are due back today, and I'm next on the list. One gets so little to eat in England, and after all it's a peculiar sort of a place. But I won't have to worry about early morning patrols which is something.

We're having a lot of fellows from one of the American squadrons over for supper this evening so I'm going to wash my face and brush my hair—'n everything
  'Bye, lover,
  Bo

9. AMIENS AND THE SOMME OFFENSIVE
August 1918

On Monday, July 29, Bogart Rogers began his two-week leave with a ride to Paris in the major's car. Unknown to him, the British were preparing an attack against Amiens that would launch 1900 British and French aircraft against the Germans in the largest air operation of the war. In the afternoon of August 3, the Thirty-second Squadron moved from Touquin in the French sector back north to Bellevue in the British sector, seven miles from Doullens on the road to Arras.

In the predawn hours of August 8, the British opened the Battle of Amiens under cover of a ground fog, catching the Germans by surprise. From August 8 to August 14, the RAF suffered heavy losses while bombing bridges across the Somme. During the first two days of the offensive, more than 140 planes were lost.

. . . .

Regent Palace Hotel
Piccadilly Circus
London,
August 1, 1918

Dearest Izzy

The day before leaving France a whole bundle of letters came from you, and yesterday I found a lot more at Cox's. The candy was really nice and fresh and very good. Somehow nougat seems to be much more deliquescent—or is it efflorescent?—than fudge and doesn't show the wear and tear of the voyage as fudge does.

At present I'm waiting for Bill Leaf, a fellow who went up to Canada from the south with me. He phoned this morning and is in town for a day or so. But we're way ahead of the story.

It was Monday morning that my leave came thru, a warrant made out as far as Scotland and a book of meat, sugar, and all the other necessary

Amiens from 12,000 feet, with the cathedral marked by the arrow to the left.

cards. The C.O. said his car could take me into Paris so away I went in a hurry. In Paris the RTO, Railroad Transportation Officer, said it wouldn't do any good to leave, as the warrant called for embarkation on Wednesday, and I couldn't leave France until then. So I looked up Bill Reagan.

Bill works for the Air Service, lives in the Latin quarter, and has a Fiat, and a driver to take him around. Also he speaks French. We had supper with a couple of other fellows from California and then went down to Bill's place and talked. The next day we rode all over the town, and I got a good look at it—Champs Elysees, Eiffel Tower and a lot of the suburbs.

The ride from Paris to Boulogne was an all night affair in a day coach with frequent stops and not much sleep. We crossed early, got into town about one, and I dashed away for my old standby, the National Hotel. It's not a particularly impressive place, but it's quiet; there is a plunge; and every one knows me there. Somehow or other it seemed almost no time since I'd left London.

I dashed down to see Brother Cox and find out how rich I might be, then ordered a new uniform, a nice light weight, light colored one, and went back to the hotel.

This morning we wandered out foot loose and fancy free and met any number of fellows we'd known before. Finally I met a fellow who knew Bill Leaf and who said he was in town. Later Bill phoned. He should have been here some time ago, but he's always late.

Oh yes. The day before I left the official communiqué said, "Lt. B. Rogers, 32 Squadron, while in a general engagement with various E.A. (enemy aircraft) dived on the tail of one E.A. scout firing two bursts of 50 rounds at close range. E.A. went down out of control. Confirmed by Lts. Paskill and Callender." That's confirmation, lover. . . .

'Bye dear, and a whole armful of love,
Bo

. . . .

> National Hotel
> Russell Square
> London,
> August 2, 1918

Dear Child

You're a darling, Isabelle, and there are times when I wish I might be able to forget you completely for it's so darn hard to think of you and yet feel so terribly far away and sort of helpless. You see, another large box from Albany came today, a lovely one not mussed up at all and the candy quite fresh. And what if it does make me sick—what if it does, huh? If eating such candy causes a slight indisposition I'll suffer in silence. But it won't.

Last night Bill Leaf and I had dinner at the Savoy and afterwards sat around and just talked. When you haven't seen a fellow for months there is always plenty to talk about. Bill told me lots of funny things among them the story of how, when Tom Whitman had rather a bad crash in training, he wrote Tom's sister a long and sad letter relating how Tom had been fatally injured and what a fine boy he was. Naturally it caused a horrible sensation.

This morning I slept late and about noon hopped a train for Hounslow to call on Harry Jackson. Jack was busy playing poker but we retired to his tent and had much conversation. He is waiting to be called overseas and says his chief duties each day are shaving once and eating three meals.

It's fine the thing you're doing to help feed the troops, Izzy. They appreciate things like that a lot more than you know. When every leave train comes in from Folkestone there is a buffet lunch waiting for the men who are going thru to some other place. The YMCA runs a lot of fine places, too.

I'm glad you mentioned the women, Isabelle, and to know that you can see those things as they really are and still not regard them as unmentionable. It's very true that Paris and London, and most of the cities in fact, aren't just what they should be. At the front there's no trouble but in the cities it's different. A great many fellows come on leave, don't know anyone, haven't any particular interests, and get into trouble. It isn't hard to do. But Izzy, if war doesn't do one other thing it teaches you the real worth of a decent woman. I honestly believe that there are very few fellows who don't prefer decent girls to those who aren't. It's just that when these fellows are here and don't know anyone the decent girls aren't as easy to meet. And the more I see of it all, the more I thank God that you're waiting for me up in your little corner of Oregon, dear. I'll promise you that you needn't worry for your love is the one thing that can and does keep me from doing things that I shouldn't.

The war's looking up, Izzy. This Château Thierry show may not be particularly decisive but it's the first good push our people have staged in many months and it has given the Hun something to cogitate on. It spoils his plans for a Paris drive and may even stop any other offensive. This summer is his last offensive summer. But it's depressing to think of what a wicked defensive fight Germany can put up when the organized lines of defense are reached—the lines he's been preparing for four years. Somebody's going to have to spill a little blood.

Izzy dear, I do love you and I'm not forgetting for a minute. You see, it's quite a job for me trying to really be worthy of you, dear, in fact being really good enough for you is sort of impossible.

'Night,
Bo

. . . .

American YMCA
Washington Inn
for Officers
St. James's Square
London,
August 6, 1918

Dearest Izzy

Even if you do live in Oregon where it does quite a bit of raining, I'd like to have you here to see what real rain is. It's done nothing but rain for the last three days.

Saturday night Jack came in, we had dinner, and went to [a] very good show. "The Maid of the Mountains" which has been running for nearly two years. The music is very pretty and Jose Collins who sings most of it can surely put it over. Reminds me of the days when I was a would-be impressario and you were my leading lady. No longer am I an impressario, but you're still my leading lady.

Sunday I went back to Hounslow with Jackson and watched the flying machines. Yesterday was a holiday—Bank Holiday whatever that is—and I went out to Northolt to see Bill Leaf. . . .

At present I'm waiting for a chap named McBeau who came over a few days after I did. . . .

McBeau and I are going to see an exhibition of naval photographs. The Ministry of Information gets these different exhibits up. They had one of war photographs a couple of months ago that was wonderful—great colored photographs taken on all the fronts and most of them of real action.

Much, much love from
Bo

. . . .

American YMCA
Washington Inn
for Officers
St. James's Square,
London,
August 10, 1918

'Lo Lover

. . . All my plans for getting out of this benighted village have gone flooie. I've met a lot of boys I know, found several good places to eat, hate to travel alone, sleep late, eat much, and am nearly broke from buying theater tickets. Really tho, Izzy, I feel heaps better than I did—the change seems to make a lot of difference.

Last evening Jackson blew in and announced that he was going to France this morning. We had quite a long talk and then dinner. I wish he was coming to the squadron, but he's flying a different type machine, and it can't be done.

I've sort of passed the time doing any number of things. Went out to the hospital a couple of days ago to see one of the boys who was in the squadron. Do you remember my telling about the chap who landed and had the bomb explode under him? That's the one—his feet are all healed

Lieutenant H. H. "Harry" Jackson of the United States Air Service, whom Rogers last saw six days before his death.

but his face is still pretty bad. It's all bound up and he can't smile, smoke, or eat anything that requires much chewing and he's likely to carry a few scars around for a while.

. . . The war seems to be going very nicely from all reports. Whoever is staging this Amiens show is doing it right and getting somewhere with it. I'd like to know where the squadron is and what they are doing and if my old chariot is still working. They're probably ground strafing in this new show. We haven't missed a push this summer, so I don't see why we should now. I noticed that yesterday's air official gave fifty British machines missing, so someone must be having a sticky job. Fifty is a good many for one day especially when our people are pushing.

The war is a long way from over, Isabelle, but it looks as if [David] Lloyd George said it when he said the Hun's chance was gone forever. It's something to know that all real danger is over and the tide has turned.

There's a dance here tonight—very annoying music and beautiful?? English girls all over the place. They can't dance, Izzy, and are all so simple looking.

Lots and lots of love,
Bo

. . . .

>Washington Inn
>for Officers
>St. James's Square
>London
>August 13, 1918

Dearest Id

If your mother says "Id" why shouldn't I, huh, why shouldn't I? Very well, in the future it shall be used. All of which is neither here nor there.

Tomorrow at seven thirty A.M. the vacation ends, Rogers climbs aboard the boat train at Victoria and goes back to work. And I'm not sorry, Izzy. It's been a nice leave and done me a world of good, but I'll be glad to get back and see what's been going on. One almost gets to like France after having been away for a while.

This morning the lady at Cox's gave me three lovely letters from you and one from your father to say nothing of a box—apparently the fig pastes. Not having any too much room in my bag—good English expression "bag"—I had them sent on to France. They'll taste ever so much better out there anyway.

I'm enclosing a few pictures. There's some kind of an order in effect forbidding the sending of pictures with machines for a background. Bill Leaf sent a lot and was dragged up before the brigade for it.

Mailed a scarf with a border of RFC regimental colors. They used to have them all in the blues and red, but you can't get them any more....

Allan Crary and I had a lot to talk about. Just about everyone is in the army. His brother George is out in France with a regiment from Camp Lewis, Bob Lytle, who you may have known, is out there somewhere, and everyone else probably will be shortly. It's a funny old war. Just as Allen said, who'd have thot that we'd meet over here in England.

Must pack, lover dear, and get up early in the morning.

Good night, Izzy,

Bo

. . . .

[Boulogne]
E.F.C.
(Expeditionary
Force Canteen)
August 14, 1918

Dear Child

Here's me back in those dear France again and waiting for transport to come and take me back to the squadron. There is no train to a town near where the squadron is now located until tomorrow morning but I found a tender from a squadron down that way and am going down with it after dinner.

Had a fine trip across the channel. The weather was quite Californian and the water like glass but just to let the folks know that the war is still going there were destroyers on each side of us and a couple of "blimps" overhead.

This is a great old town, smells fishy and dirty, looks dirty, in fact it *is* dirty. It's a busy place and crowded with soldiers and transport of all kinds. The harbor is filled with ships, and the trains are continually coming and going. It's a real war time village.

I've been wandering around a bit, walked up to the old town, which is up on a hill above the harbor and surrounded by a great stone wall. Inside are the Hotel de Ville, the Cathedral, and any number of ancient and interesting old houses and buildings....

Bo

. . . .

[Bellevue]
32 Squadron
France,
August 16, 1918

Dear Lover

My, My!! What an awful place this France is, so noisy and tumultuous and unpeaceful. My poor old nerves aren't quite up to all the excitement yet, and I nearly threw a faint last night when Fritz came over and scattered a few pills around the place. But as usual I'm miles ahead of the story.

First of all I finally got to the squadron after riding from Boulogne in a tender from another squadron to a town very definite. It was late when we rolled into said town, so I stayed all night at the hotel and phoned the squadron in the morning. They sent a tender, and I arrived here in the afternoon.

We are much nearer the line than we've ever been before, rather too close for comfort. We hear a good deal of the noise from the guns. Also at night we can see parachute flares which are sent up from the front line.

Of course a good many things have happened. First there was the move, apparently a good one to have missed for it rained just as they started and things were in an awful mess.

Just as I suspected the squadron had a rather sticky time of it—lost four men and had any number of scraps and exciting times. Poor old Rube Paskill went west in a big dog fight with a lot of Huns. The others all were put out in scraps, too. Some new pilot took my grand old aeroplane out, ran short of petrol, tried to put it down in a field and washed it out. The mechanics were all for murder, and I wasn't particularly pleased, for it was one of the oldest and best machines in the squadron.

Several fellows have knocked over a Hun or two and done a lot of good ground strafing shows. Then of course on the bright side Simpson and Green both got the Croix de Guerre and Palm for the shows with the French. They had a large parade, French generals and aides-de-camp, all sorts of ceremony and nice new medals. Then three of the new pilots are very good, have all been instructors in Canada, can surely fly, and are regular fellows besides.

Yesterday afternoon I pulled myself together, girded up my loins, and went out to fly. My new bus not being ready I took Simpson's and nearly threw a dozen deaths. Seems I'd forgotten completely how to fly. First when I was tootling quietly and peacefully around thinking about what a nice place England was some enthusiastic person got on my tail and I couldn't get him off being unable to think of anything to do except climbing turns. Finally I got up enough courage to do a half roll which did the work, but about two minutes later another person picked on me so I landed very slowly and nearly ruined the machine. This morning I've been working on the new kite and am going to take it up later.

Last evening after dinner the lights went out, the klaxons blew long and loud, and when the noise stopped we could hear the peculiar drone of a couple of Huns. They circled around for awhile then suddenly dropped some parachute flares. Fortunately they weren't right over us, for they light the whole place up like day. They headed for a nearby town and prepared to lay their eggs. The usual Fourth of July celebration followed, first the searchlights, five or six of them feeling around for the Hun, then flaming onions and streams of tracer machine gun bullets from the ground. Fritz dropped another flare and laid his eggs. Then the searchlights got him and a perfect whirlwind of stuff started up, archie, onions, machine gun stuff, rocket flares, everything. The Hun started home, finally got out of the light, and very shortly thereafter all was quiet. It seems they are over here almost every night. Only a couple of nights ago one of our night flying scouts got one down in flames. Anyway archie gives them a merry time, and they don't do much damage.

This is a very fine aerodrome, large and flat, the hangars of sheet iron, and we live in corrugated iron huts, nice large places with good floors, but they do get warm during the day. The mess is a large wooden hut with a fireplace and many beautiful pictures, mostly collected from the "La Vie Parisienne," adorning the walls. It ought to be a fine place in which to spend the winter, but we'll probably move again shortly.

The fig pastes came this morning. They were great. A large part of them have gone missing due to the combined efforts of several other hungries. The little cherry things were good, too.

Guess I'll stop now, dear, and see about my kite. Don't forget, Isabelle, how I love you.

All yours,
Bo

. . . .

[Bellevue]
France,
August 18, 1918

Dear Child

Hasn't been an awful lot to write about since the last letter—almost nothing in fact. Simpson is going home in the morning, probably to get

Rogers taking off from the Bellevue airfield.

a squadron of his own in England, and we hurled a large farewell dinner for him last night. For guests we had several friends of his from other squadrons, for food we had just about everything and plenty of it, and for refreshments—there was plenty for refreshments. It was quite a good party and ended up in the sergeant's mess where Simmie insisted on going and bidding them a fond farewell. I hate to see him go, for he's one of the finest chaps I ever knew and a wonder to follow. However he's been out here three years in the flying corps and one as a Tommy. He deserves a squadron if he does get one which is very likely.

The weather has been bad, and there hasn't been a patrol since I came back. I had my new kite up yesterday. It's rigged beautifully and light as a feather on the controls. Altho the engine is new and a bit rough it gives plenty of power and speed. I suppose one machine is just about as good as another, but you become attached to one after you've had it for a long time, and it hasn't ever let you down.

I love you,
Bo

. . . .

Arras, fourteen miles southwest of the Bellevue airfield, as viewed from the cockpit of an SE-5A.

[Bellevue]
France,
August 21, 1918

Dearest

Wonder what you're doing this wonderful evening? I've just come in from a machine gun pit, umpteen Huns up above, archie banging away at them, pill dropping here and there, and our night bombers going to and from the line. Such a perfect night and all they use it for out here is to drop bombs on each other. Personally I think it would be a bit nicer to have you beside me, maybe singing a song. Damn it!! All I had beside me tonight was a Lewis gun and a pile of sandbags.

However the war's the thing, just now at any rate, and our night fliers are working overtime. Any number of them have buzzed by overhead going to the line. Some of them were low enough to see quite plainly, others were only one small light sliding across the sky. There was a Hun or so over, too, but they did very little damage.

This afternoon I did my first show since coming back, an escort well over the line, but nothing happened. The visibility was wonderful. From 15,000 England was very plain, while the channel sparkled like gold.

This morning our people staged a small push [Battle of Bapaume] and got on quite nicely with it. The guns had been fairly active all night, but about five this morning they cut loose with a horrible bombardment, short but very intense. Since then the guns have been fairly quiet, but the advance has worked out well and our people have made considerable gains.

This evening a lot of infantry marched by to take up their place in the line. They had a band and everyone was singing and shouting and seemed to be full of pep. I can't say I envy them their job.

Oh yes. You must know about my whiskers. A day or so ago Green and I decided to cease shaving our respective and respected upper lips and attempt to cultivate thereupon a bumper crop of whiskers, said whiskers later to be trained into neat and artistic mustachios. Us boys must be amused, besides, Green's is blondish and doesn't show up very well so I am able to gloat.

A couple of nights ago I ate rather freely of some canned lobster before going to bed and what do you suppose happened? I dreamed about you. Last night it worked again. Tonight I'm forgoing the pleasure for the sake of a tortured digestion but tomorrow I intend to recklessly indulge again. I'll bet you don't believe a word of that, but it's true.

We have a very early show in the morning, dear, and it's quite late now so just

good night,
Bo

. . . .

[Bellevue]
France,
August 23, 1918

Izzy Dear

Just a sort of hit and miss note as we're due for a show in a few minutes, going to take some bombers across and see that nobody fools around with them while they are doing their dirty work. We took a batch over this morning, a long way over, too, and had a touch of excitement. Just before they reached their objective four Huns came up under them and were about to do something mean when Green and his flight tumbled down. Green got one in flames, and the others went down in a hurry. Coming back seven Huns came up from the south and played around very politely to one side of us. We might have tumbled down and had a scrap, but we had to see the bombers to the line so didn't. Finally two of the Huns pulled up their noses and did a little shooting at us. If we had stayed for a scrap we'd have been in for a bad time for just as we were about to the line some twenty more Huns appeared above us and in the sun. The sun is awful early in the morning, absolutely blinding.

There has been a battle going on all day. From the latest dope our people are getting along nicely. The guns have been at it hard and heavy since early morning. As the Hun is indulging in a little counter battery work the noise is pretty bad.

Last night we didn't get much sleep as the night bombers were all over the place. I'd have slept thru the whole show, but we now have an archie battery right in our back yard so to speak. I'd just dozed off nicely when these people cut loose on a Hun and made an awful noise. After that we were up for an hour or so watching the fireworks and praying that no pills would drop on us.

Izzy dear, I can't have you out here but I can have your letters and they're everything. I just sort of exist from noon one day when the mail comes until noon the next day when the next mail comes.

Bo

. . . .

[Bellevue]
32 Squadron, France
August 25, 1918

Dearest

Yesterday we had no shows, but this morning we made up for it when we did a close escort for bombers so far over the line that I shiver every time I think about it. There weren't any Huns about but there was an enormous amount of archie and much of it was very good, kept us on the move dodging it, and the bombers were simply plastered.

This afternoon we're doing a nastier show. I can't go into details but it's a retaliation affair and worth the effort involved. However, we get the easiest part of the whole show.

A letter from Simpson reminds me of something very funny. Simmie wasn't going directly home to London so he sent his small pup, "Peter," home by one of the fellows who was going on leave. He borrowed a hand grip, tucked Peter into it, and was going to deliver him to Simpson's folks in London. But in Boulogne something happened. Peter was last seen by a hotel porter dashing down the street. Simmie asked what had become of his dog, so I sent him this chap's letter telling of the mysterious disappearance. Too bad. Simmie and Peter were great friends.

My new flight commander came out a day or so ago and has turned out to be quite a decent chap. His name is Zink, he's very young, and hasn't been out here since last December when he was shot down. He's a good pilot and very satisfactory. It's a relief to get decent fellows.

During this bum weather I've been doing a lot of reading trying to improve the mind a bit. J. M. Barrie's stuff is good. Also a good tale by Mrs. Humphry Ward, "Missing." It's quite worth reading.

I love you, dear.

Bo

. . . .

[Bellevue]
32 Squadron,
August 27, 1918

Hello

The pictures came. They're great Izzy—I've been looking at them about every ten minutes and sort of saying to myself "Look what I've got."

This morning at eight I slumbered peacefully, as our early show had been called off. A vague but familiar voice spoiled everything. After many rubbings of the eyes I discovered the owner of the disturbing voice, a chap named Crabb who trained at Stockbridge when I was there and who is out here in the same squadron with Shapard. He had been out on a patrol and made a forced landing in a field a few miles from here. He was around for a couple of hours—had breakfast with us, and finally departed with a couple of mechanics. He hasn't come back yet so I guess everything must have been fixed up.

Last night I received quite a shock, maybe not as bad as it might have been for I'm getting used to them. Had a letter from Harry Jackson a day or so ago telling where he was, so last night four of us went over for dinner. Jack wasn't there, went missing in the afternoon. It seems his squadron had done a patrol in the afternoon, ran into a big bunch of Huns, and had a bad fight. The wind was strong and blowing right into Hunland. Since they were pretty far over things look rather bad. Of course he may not have been able to get back and is a prisoner, or he may have just got back and been unable to get word thru to his squadron. I hope he's alright, but as some other fellows were missing too, it's a bit doubtful.

It's tough to see things happen, but it can't be helped. I only wish it might be different. I heard a day or so ago that Bill Taylor was shot down in flames.

The show I mentioned in the last letter, the little personal affair we had to settle with the Imperial German Flying Corps came off very nicely, altho it was a bit exciting. We did our part of the show as per schedule and had to fight to do it. For half an hour we sat over several of the largest Hun aerodromes. The sky was full of Huns. We got two of them and didn't lose anyone. In fact there were only two machines missing in the whole show, and one of them had a forced landing on the other side. The Germans haven't much to say in the air. They are still hard propositions to handle, and we may lose quite a few machines, but a hundred of our machines cross the line to one of theirs. That makes a vast amount of difference in artillery observation, reconnaissance, photography, and bombing.

The war seems to be going slowly, but very steadily in our favor. Every day there is some gain, often a good size one. From all reports our casualties are quite small. Winter is just about here and I imagine things will slow up some what then. But next spring—there will be much, much war next spring. . . .

Your mother writes dandy letters, Id, calls me Bo 'n everything.

Your letters are so sweet, Isabelle, and they mean so very much to me—more than you will ever realize.

All yours, dear.

Bo

. . . .

[Bellevue]
32 Squadron, France
August 29, 1918

Dearest Child

About two minutes after I had finished the last letter to you we got a tender and rode up to the line. At least we were about two kilos from the front line and on ground that had been taken from the Boche only a couple of days before. The roads were of course very bad, so we went overland toward the line. There were dozens and dozens of things to see, and it was all very interesting. We were in several villages—piles of rubbish—that the Hun had occupied. The whole place is pretty messy, shell holes, trenches, barbwire, dugouts, mine craters, and plenty of mud making walking a bit difficult. The ground was strewn with all the stuff that is left behind in a retreat, helmets, rifles, kit of all sort, and any amount of nondescript articles. In one place the advance had been so recent that corpses were still about. We found all sorts of curious affairs. Some of the fellows carted a lot of it home with them. However it really doesn't pay to bring it back as most of it can't be sent out of France, and it's a bother to carry it around. Besides, there will be plenty of souvenirs before this mess is over. One infantry officer told us that to get anything really valuable you had to be about ten yards behind the men going over.

In a deep railway cut, there were some wonderful dugouts and a great deal of material, large piles of shells, hand grenades, and trench mortars. The Hun must have moved out in a hurry, as very little had been destroyed.

Coming home a battery of heavies cut loose about twenty yards from the road as we were passing. Such a noise I never heard before, first a titanic bang, a flash of fire and a cloud of smoke, then the whine of the shell getting fainter and fainter, and finally an almost inaudible "wonk" from the other end. And just under the muzzles of the guns was the men's mess where several of them were eating in sort of bored fashion. The infantry and artillery, and tanks, too, can all have their jobs. I like mine. At least back here we can eat without feeling that maybe Fritz will send one over and upset the food, and there's not much danger of our beds being suddenly covered up by five feet of mud. Those men up in the line must have nerves of steel. I'd go crazy as a March hare if I had to be up there one whole day.

Yesterday afternoon Shapard sauntered in. He had come down with a working party to get Crabb's machine which wouldn't fly after all.

I was glad to see him for we got to be pretty good friends in England. Shap has one Hun, rather a funny affair. As Shap chased him, the Hun apparently had a colossal wind up, dived very steeply and never pulled out, just dived right into the ground. So Shap got credit for him without firing a shot.

Lots and lots of love, dear,
Bo

10. CAMBRAI AND THE HINDENBURG LINE
September 1918

During September, the western front moved toward the Hindenburg Line, an impenetrable maze of barbed wire, trenches, canals, concrete shelters, and armored machine-gun emplacements. If the Allies could breach the line, four years of trench warfare would end, because there was open country beyond.

The RAF Ninth Brigade consisted of thirteen bombing and fighting squadrons, including the Thirty-second. During the early weeks of September, bombers from the Ninth Brigade hammered rail junctions while fighters flew escort and offensive patrols to break up massed flights of twenty to forty enemy machines. Bad weather hampered flying for more than half the month. Weather permitting, squadron pilots engaged in dogfights over Douai, Cambrai, Le Catelet, Bellicourt, and as far south as Saint-Quentin.

At the end of the month, on September 27, the Battle of Cambrai and the Hindenburg Line opened in the predawn darkness. Nearly 1000 aircraft supported the British First, Third, and Fourth Armies as they advanced on a line from Saint-Quentin to the Sensee River north of Cambrai.

. . . .

[Bellevue]
France,
September 1, 1918

My Dearest

The day's labor was strenuous, even a bit exciting, in fact some of Kaiser Bill's hired men threw quite a scare into us. The wind is from the west and at 15,000 feet too strong for all practical purposes. We took a batch of bombers over to a place they should have never gone on a day like this. So far so good. The bombers turned for home and came like the very dickens, so fast that we could just keep up with them. Our flight was doing the close protection, and the two other flights sat up above, but unfortu-

Pilots of the Thirty-second Squadron: Carson, Wilderspin, Amory, Callender, Russell, Spicer, Tancock, Farquer, and Farson.

nately got so far behind they weren't the least use. There you have it: five of us protecting the bombers, the rest of our machines specks in the distance, and nobody making much headway against the wind. About fifteen miles from the line enter seven Fokker Biplanes, very good machines and very good pilots. They were between us and the line, also in the sun. They tumbled down. We didn't want to fight as we barely had enough petrol to get us back, but we had to fight. The bombers kept serenely on their way while we all went around and around, up and down with many Fokkers chasing us and every body shooting. Two little devils got on my tail and I did things that I never suspected could be done with an aeroplane. Finally, one left me, and a good burst drove the other one away. By that time Zink, my flight commander, was way below, two Huns after him and last seen entering a large hospitable looking cloud. Two of us crossed the line with the bombers. So we came home, told our tale of woe, and said we thot the others were in Hunland. About ten minutes later in comes Zink, his machine full of holes, but very glad to be home. His two Huns chased him into a cloud, he tried to dodge from cloud to cloud, but one Hun found him, and they fought right down to the ground, Zink finally getting the Hun, and coming home just at the tree tops. Two still missing from our flight and three from another. One of our fellows turned up about an hour later. He had landed at an advanced aerodrome with the other chap who was wounded in the arm and his machine riddled.

Three still out. One came home a few minutes ago. He had gone down over a town about fifteen miles in Hunland thinking it was a town just on our side of the line. They began to archie him, he couldn't climb without making very little speed west owning to the wind, so he went down to fifty feet and hopped the hedges. Just before he reached the line he saw a Hun single seater tootling peacefully along at 500 feet (suspecting nothing like someone coming up under him from the east) got right on top of him and finished him off. Whereupon he chuckled sinfully to himself, crossed the line, landed to find out where he was, and came home. Two fellows are still missing altho they may show up yet. Not a bad afternoon for thrills.

Yesterday I had one or two small sensations myself. A chap named [Monte] Tancock, he of the episode above, and I went out to see if there might be a playful Hun or so about. There were, but they wanted to play rough. Being new, Tancock went down on a single Boche when there were five more up above. I went down to bring him back, as I was sure he hadn't seen those above, and the five put their noses down and came after us. Tanny didn't see the Huns until they started shooting. It didn't take him long to clear off. One of them got on my tail whereupon I dived. He dived. I pulled out and dived again, he pulled out and dived again. I half rolled, he half rolled. I was going east then, so half rolled again, and he was still there. I was getting a bit nervous, so put my nose straight down and dived like I've never dived before. I pulled out gradually, going west all the time, and he finally turned off as we were nearing the line, and they won't follow you across the line. Anyway that was enough for a day or so. I'll tell you all about Tancock one of these days. He's an interesting person.

If you'll glance at the upper right hand corner of page one you'll see that September is here and between you and me, lover, it's here in more ways than one. Today I wore a pair of silk gloves, a pair of chamois gloves, and a heavy pair of flying gloves over the lot, and my hands nearly dropped off. The winds are getting cold; the nights are getting cold; and the sweaters of Miss Young are about to see active service again. The socks see a good deal of service. Something tells me that France is no winter resort.

Ho hum! Lovely war, but you keep an eye on this war. There are going to be some big surprises one of these days. Also—

I love you, Izzy.

Bo

. . . .

Lawson, Zink, and Green at the Bellevue airfield, September 1918.

A plane flown by Lieutenant A. H. "Sandy" Sandys-Winsch nosed over, a common occurrence on rough landing fields.

[Bellevue]
France,
September 3, 1918

Lover Dear

This is probably one of the earliest written letters you've ever received.

We were dragged out at a quarter to six, then after we had sleepily swallowed the customary egg, toast, and tea the weather started getting very poor all of a sudden and now we're just waiting around for something to happen. Half of the fellows have gone back to bed, some of the others are enjoying a small poker game, as for Rogers, up to the old tricks again.

It's as cold as the dickens these mornings. Never did getting up for eight o'clocks [classes] cause me the intense agony, the acute suffering that crawling out for a six o'clock does here. Fact is and I'm getting disgustingly lazy and unambitious.

Yesterday was another dud one, but towards evening four of us managed to wake up sufficiently to go looking for stray Huns. There were no stray Huns, in fact about all the Huns we've seen in some time have been in well organized and pugnacious bunches. I would that one might run across a fat, slow and comfortable two-seater now and again but they seem to be pretty scarce.

Last night while we were at dinner who should wander in but the chap who was wounded Sunday [A. H. Sandys-Winsch]. He had been in a French hospital but managed to get back here to collect his kit. He looked sort of wild eyed and very much unshaven and very wrathfully declared that no gentlemanly Hun would sneak up behind a man and shoot him while he wasn't looking.

That's all—'cept a little love.

Bo

. . . .

[Bellevue]
France,
September 4, 1918

Isabelle Dear

Everything's all wrong tonight and the only thing left is to take it out on you. You don't mind if I unload a trouble or so on you, do you, lover? If I can write them to you it's almost the same as being able to tell them to you.

It started yesterday afternoon when in one of the easiest shows we've had in many a day, poor little Jerry Flynn was shot down in flames. A bunch of Huns came down on his flight, and before we could get to him he was gone. Then there was a wicked dogfight everybody getting more or less shot up. Jerry was only a kid and about the most popular person in the squadron. Everyone was pretty much broken up over it, more so than I've ever seen them before.*

---

*In an article that appeared in *Popular Aviation* (published today as *Flying Magazine*), Rogers described what happened after Flynn was killed:

Callousness and a hard-boiled and unsympathetic attitude were the chief salvation of the air service.

Outwardly nobody was sympathetic. They had feelings, of course, but not very obvious ones. If your best friend was shot down you masked a breaking heart by declaring he was a damn fool who should have had better sense. You tried to make the rest of the boys believe he got just what was coming to a fellow who made a silly mistake—and they pretended to agree. It was his fault and it served him right!

Only once did I see an entire squadron really go to pieces. It was a most unpleasant affair.

Little Jerry Flynn was the pet of the squadron. "Gee Whiz," they called him—a Canadian kid who was just eighteen, but a captain, a flight commander and a veteran at this new kind of slaughter. The boys didn't like Jerry—they loved him—as you'd love a kid brother. When he would start away on a patrol with his head barely peeking over the edge of the cockpit he looked like a small child—he had little boy eyes and not the sign of a whisker.

One afternoon the whole outfit went on a short patrol over the lines, so short and

Last night Green, who was Jerry's best friend (they've been together for ten months) went all to pieces, nerves simply gone. It gets to you to see a boy go like that, for while Green is an old man in the war, he still has his twentieth birthday to celebrate. He's been out here too long, nearly ten months, and will probably go home almost any day now. The C.O. told me today that I was to be recommended for his flight, but I doubt if it will go thru. I haven't been out here long enough and haven't done a thing to deserve it.

There are only three flying officers in the squadron who were here when I came. Makes one feel pretty old and experienced. It surely is hell to see them pass by, Id. But the only way to do is simply to forget that you ever possessed such a thing as an emotion or a nerve and carry on just as if nothing had happened. Is it any wonder that fellows go to pieces?

Good night, lover. I always wish terribly that I might be able to say it some other way than with a pen, even without any words at all. But—c'est la guerre.

Love

Bo

---

so simple that everyone got careless. Three miles over, a dozen Fokkers appeared from nowhere. Jerry never saw them. Before the rest could come to the rescue he was tumbling down in flames.

His particular pal was a chap named Green. They were like brothers—lived together and played together. When Green landed and crawled out of his plane the tears were streaming down his cheeks. He'd been crying all the way home.

"Poor little Jerry," he sobbed. "Oh, my God! they got him in flames! It was all my fault—I let him down—I didn't protect him!" He wailed out more and more worse things than that. If you've ever seen a football player whose mistake has cost his team an important game, you may have a faint conception of how this lad talked and acted. His heart was broken.

Everyone was on the verge of tears. It was too great a tragedy to conceal. It penetrated their calloused exteriors and jabbed at their hearts.

It was at dinner that the whole outfit blew up like a toy balloon. There were more cocktails than usual—it was the easiest way to forget. But for once they didn't work that way—they only made matters worse.

By the time the soup appeared everyone was three sheets to the wind but the teetotalers. There were guests for dinner and champagne was in order on guest nights. They guzzled it like water but it had no effect—the place was a morgue. Nobody talked —nobody dared talk. Jerry's seat was vacant—there wasn't a soul who would occupy it.

It happened suddenly.

A kid named John Trusler grabbed his champagne glass, hurled it the length of the mess, leaped to his feet and started a vivid impersonation of a lunatic passing his entrance exams for Mattewan. He swore and cursed and cried. He cursed God and

. . . .

[Bellevue]
France,
September 6, 1918

More darn fun, Id. Been flying practically all day and maybe got two Huns, surely one and quite likely another. The funny part of it is that I could have had another if both my guns hadn't gone flooie.

It started out this morning. We had no regular show so I went up about ten o'clock on a still hunt for two seaters. Had an idea where there might be a few. However these particular Huns come over very high so I went up to 20,000 and sat there in the sun, but nearly frozen. Very shortly along comes a nice Hun two seater, fat, slow, and comfortable, and about to take a few pictures. Right away I got over anxious and scared him east. He started to climb, and I started to climb. When he came over again I was at 21,000, and the Boche a bit below. I waited until he was where he could do the least harm, then went down under his tail. If you get under a two seater's tail he can't shoot at you because the tail is in the way. He turned east. I turned and got right under him, then did the dirty work. When last seen Fritz was doing a funny spin and pouring out white smoke. I couldn't

---

the Germans. He cursed the war and the army. He cursed his parents because he was born. He told little Jerry Flynn—who he knew was in the room listening to him, who was sitting right there in that chair!—that he didn't have to worry. They couldn't kill him and get away with it.

He cursed far more capably and colorfully than a lad of his years had any right to do. They tried to stop him, and finally pulled him down in his chair. He hid his head in his hands and sobbed horribly. Half of the fellows were bawling. The rest were trying to quiet things down, but feeling no better than the weepers.

Finally it subsided. The orderlies brought more wine and the original objective of finding solace in alcoholic oblivion was successfully attained.

The next morning the hard-boiled masks were up again. They had to be, of course. They all told each other it was just another binge. They kidded Trusler for getting on a crying jag.

It was a tough break, they said, and the place wouldn't be the same without little "Gee Whiz"—but, after all, "C'est la guerre," and he had nobody but himself to blame. Why in hell didn't he watch his tail? Any guy who let himself be caught napping couldn't get sore if they shot him down!

Green had apparently regained his bravado with the rest. He went on an early morning patrol and evened things up for Jerry by knocking a Hun down in flames.

But later in the morning he collapsed—suddenly, unexpectedly and completely. His nerves snapped with the twang of a broken flying wire and they sent him home for a long rest. ("The Startling Truth About War Fliers," *Popular Aviation*, December 1930.)

Captains Flynn, Simpson (with Peter), and Green behind Major Russell.

see him crash as there were clouds a few thousand feet below and he disappeared into them.

A little later along comes another. In the first affair, my Lewis gun broke and left me with no top gun. This second bird was slightly above and stayed there. He started turning, so the observer could shoot. I kept doing larger circles to keep behind him. I could see the observer shooting now and then but always wider. Finally I got tired of fooling around, pulled my nose up, and pressed the lever. About five shots and the gun stopped. It was a separated case, something that can't be fixed in the air, so home I came.

This afternoon we escorted bombers. Over their objective six Huns came up and started shooting at very long range. We tumbled down, and I got a good burst at one, but was pretty sure I'd missed. However when we came home the bombers reported a Hun down in flames, so maybe I'm to blame.* Too bad.

Then before dinner the C.O. asked me to take up his bus and see what was wrong with it. He couldn't get it to run right. Feeling quite puffed up I took it up, and strangely enough discovered the trouble.

Wholesale love, Izzy dear.

Bo

. . . .

[Bellevue]
France,
September 9, 1918

Izzy Dear

Nice rainy day, lots of ink to throw, and quite a few things to write about so what is there to complain of?

Glad to hear the picture frame finally showed up. One never knows what will happen to things sent from out here. I'm sending your folks an ash tray made from the base of a German archie shell case—not much to look at but a bit of a curiosity. . . .

Saturday morning I went up again looking for two seaters but it was rather cloudy and none were around. Incidentally it was very cold at 20,000 and one of my fingers is frozen or some such thing. Hands are one thing you can't keep warm and it's awful when you come down and your hands begin to thaw out. Since Saturday we've done no flying.

Last night we hurled a party. It was like this. One of the boys in the squadron, an American from Syracuse, [F. L. "Bud" Hale], ran into his old regiment who were in the line not far from here. They were all fellows he knew very well and had been on the Mexican border with. He had several of them up last night. They were a fine lot of fellows, some of the best I've met out here. These poor people in the infantry don't get an awful lot of joy out of life. As this bunch had just come out of the line, they had a better time than they would probably have had otherwise. Without throwing any sweet scented bouquets, we can fix up a very attractive looking dinner. One of these fellows said it was the most beautiful sight he'd seen in France.

---

*Two Combats in the Air reports were filed by Rogers on September 6, 1918. His claim of a Rumpler two-seater destroyed at 11:00 A.M. at 21,000 feet was confirmed to have crashed near Reisel by the Forty-second AA Battery. The second report claimed a Fokker biplane destroyed at 5:00 P.M. near Holnon, west of Saint-Quentin, at 14,000 feet. The Thirty-seventh Squadron confirmed this enemy aircraft to have burst into flames and burned up in the air.

Captain Wilfred B. "Wilf" Green, the nineteen-year-old flight commander and seven-victory ace of the Thirty-second Squadron.

Green went home yesterday, at least went to the hospital and will go home from there. I hated to see him go. He was a fine boy and a wonderful pilot, but he'd been out here too long. I was a bit premature in my predictions and probably won't get his flight. Someone will come out from England and take it. I haven't been out here long enough or had wings long enough but I'll get one yet. . . .

What else? Oh yes! Our movies are running again four nights a week. Saturday night we saw a fair Fox picture and a Keystone comedy, all taken in Los Angeles and all very familiar.

We have a new padre, or at least the wing has. He's around quite often. His name is Wilson, but he's commonly known as the "Cannibal King" for the simple reason that he was a missionary somewhere in the South Sea. You've no idea what a fine lot of fellows the padres are out here, and what a lot of good they really do. None of them are the kind of men that make you feel uncomfortable and restrained. They're all regular fellows who have seen the war, know what it is, and act accordingly. A great many of them have decorations, which they deserve for their work right up in the line, often in attacks. A non combatant's job in action isn't a very pleasant one.

The war as you must have noticed isn't at all bad. The Hun is just about back to his old Hindenburg line, beyond it in one or two small places. What will happen after that I don't know. The Hindenburg line is a wonderful affair and he ought to be able to put up a pretty fair defense there, but you never can tell.

Guess that about concludes the performance.

All yours, dear,

Bo

. . . .

[Bellevue]
On His Majesty's
Service
France,
September 11, 1918

Dear Lover

I've a mere matter of a hundred or so letters to worry about before wooing Morpheus—good phrase, that. You see, it's the old orderly officer job again.

Two more dud days and no work. Yesterday afternoon four of us went up to the line to see the war. And so we did, so we did! We drove out [on]

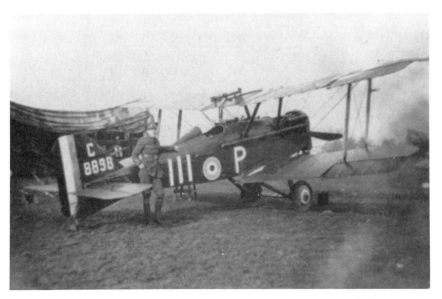

Lieutenant Evander "Shap" Shapard of the Ninety-second Squadron, and his "bus."

a road that has seen much heavy fighting in the last push, drove and drove until the traffic began to thin out and finally ceased altogether. We figured we were getting pretty close and finally stopped at a cross road to ask some Tommies just where we were. We hadn't any more than got out of the car when z-z-z-z—whang!! Then three more burst within a hundred yards or so. We hiked for a nearby dugout and learned that this particular place was under observation and getting too hot for comfort. Everyone lost enthusiasm, and we went back where it was a bit more peaceful. Again I say the infantry can have their job—I like mine.

Today we went up again, but only to get some Hun shell cases to make into souvenirs.

Now about Shapard. Several days ago a couple of fellows from his squadron landed here, said they had just had a gosh awful scrap and that Shap and a couple of others were missing. One of the fellows was certainly gone, and the last seen of Shap he was being chased by five Huns. So I figured that another good boy had gone west and thot of a few of the good parties we'd had. But it seems that the old boy fooled 'em—read about it in his letter.

An armful of love, dear.

Bo

[enclosure]
SHAP'S LETTER
E.F.C. Officers Rest House and Mess Sunday 8 Sept 1918
Dear Bo—
For fear you might come up to see me during the next two weeks I'm writing.

Last Thursday we went down south and got into a hell of a fight. Like a fool I went to the bottom and got chased from Cambrai to the line by five Fokkers. They shot the trigger group off the Lewis and made a pepper box of the rest of the machine—but never touched me—three longerons shot through—so the C.O. thinks I need a rest. Therefore two wks in Blighty.

I've appointed myself official caretaker of the top of all fights hereafter. Best of luck and lots of Huns.
Shap
P.S. I got one before they started on me.

. . . .

[Bellevue]
France,
September 16, 1918

Dearest Lover

Having done just ten minutes short of seven hours flying this beautiful day I must confess to a feeling of languor and a strong desire to hit the Ostermoor, but I'm reasonably happy and must say the day's labor has not been in vain. Got another Hun.

The morning show was early, an offensive patrol, and quite unexciting, not a Hun about the place. Zink and I nearly went over to strafe a Hun balloon, but thot better of it and didn't. Balloon strafing is a nasty job, they shove up everything but bricks from the ground as soon as you get anywhere near. One of the Hun's cute little tricks is to put up a nice fat, tempting balloon where it can be easily gotten at and incidentally containing a basket full of high explosives which they set off when you are near.

After coming home and having breakfast three of us set out to look for two seaters as it was a perfect morning, the first in several days. It took an hour to get up to 20,000. I tootled up and down the line for another hour and never saw a single Hun. By golly, it's cold and monotonous work looking for these people, and when you don't see a thing you come home completely fed up.

The afternoon show was the same as the morning one, only different. Same place and height, but much excitement. First we saw nothing, then

Lieutenant C. W. "Tommy" Thomas, a newcomer to the Thirty-second Squadron in September 1918, in a Sopwith Camel.

we saw several Boche very high up and well over. The next thing we saw was a scrap, which ended up by our machines getting chased home. We were too far below the Huns to be able to get above them and weren't particularly anxious to let them tumble down on us. They fooled around, and we fooled around. They went north and then west and tried to get into the sun, and finally when everything began to look cold, six British machines walked right in under these Huns who then tumbled down. We waited until all the Huns were below us, and the fight was on.

As soon as it started other machines—all ours—came from all directions. The place was simply cluttered up with aeroplanes. Every time you turned you nearly bumped into one, and the Huns were on the bottom of the pile. I was looking around for someone to pick on when a Hun got on the tail of one of our busses who was chasing another Hun. This particular Fritz was too interested in the man in front to worry about the man behind. A hundred rounds sent him down in a spin.* Altogether we got four out of the scrap and lost nobody.

---

*Rogers's Combats in the Air report for September 16 claims that a Fokker biplane was driven down out of control at 6:10 P.M. at 8000 feet near Sancourt (four miles northwest of Cambrai). The enemy aircraft was confirmed destroyed by the Ninth Balloon Section, Third Brigade.

Monte Tancock, whose specialty is getting shot down, lived up to his reputation and was forced to land on our side pretty close to the front line. He got fixed up, had some Tommies swing his propeller, and when he took off he caught his tail skid in a field telephone line and came home trailing about 200 feet of wire behind him.

The funniest part of the whole show involved a new chap in our flight named Thomas, a good pilot, but only doing his first shows. He and another chap were following me. When the scrap was over, and we were all split up, I looked behind and there was Thomas still trailing along. When we got home I asked him what he thot of the scrap. He said he suspected there was a fight going on, but he couldn't see anything that looked like a Hun, so he just followed me. All of which goes to prove that you learn to see all over again out here.

Great excitement. One of our night bombers came over a few minutes ago, fired off a few lights, and perched on our aerodrome. It is very moonlight but he made a nice landing for any kind of light. There were two in the machine, pilot and observer. They had been bombing, but came back because of engine trouble. Some of these night flying pilots are certainly stout men. They do all sorts of shows in any old kind of weather and a long way over the line. . . .

No. I can truthfully assert that as far as tobacco as an article of diet I'm like little Bobby Reed.

>"I do not chew tobacco"
> Said Little Bobby Reed,
>"I will not even touch the stuff,
> It is a nasty weed."

They're still hunting for the man who done that poem.

Good night, dear Isabelle.

Bo

. . . .

[Bellevue]
France,
September 18, 1918

Dearest

That all-important affair, the weather, has been nothing to write home about and we've had no show today. Yesterday we had one, but one of my magnetos went flooie and I never got off the ground. In the afternoon I took four new fellows out and showed them the line from a safe distance. After bringing them home I went down south to call on a chap named Corse, who you may have heard of before. He was with me in Texas and came out here with a British squadron some time ago, altho he is in the U.S.A.S. Unfortunately he had just been transferred to an American squadron, so my visit didn't turn out very well.

The aerodrome belonged to the Hun not very long ago and has a lot of Hun buildings on it. It is full of shell holes that have only been filled up with soft dirt making it very bumpy for landing. The country for miles around is absolutely desolate, not a tree nor a house standing, shell holes and old trenches every where and a most unpleasant odor. The nearest undamaged town is miles away. It's not much of a place to spend the winter. We're all praying we won't be sent up to another like it.

I've a lovely picture of myself and the C.O.'s dog that I'll send along in a day or so. I may as well warn you, dear, that said picture does me a grave injustice, causing my exceedingly handsome countenance to appear not at all handsome, and by some devilish trick of the camera making my nose look several sizes larger than it actually is. If your letters should suddenly cease I'll know what caused it.

Had a letter from Simmie in England. I wish he was out here again, for I've yet to meet a fellow I liked more. Some of these new chaps we have are alright, but they don't seem to improve much on acquaintance. When it comes to flying, they aren't in the same class with Simpson. He wasn't particularly brilliant, but he knew more about machines, saw more, and used more judgment than anyone we have out here now. Also he was very easy to follow and very useful when it came to shooting Huns off of your tail.

This afternoon a South African chap named [Bruce] Lawson and I went for a long walk over the fields and found a big berry patch where we stopped long enough to fill up on blackberries. All the farmers, mostly women, were busy taking in the grain.

Sometimes I can hardly believe that you love me, Izzy dear, and are waiting for me to come back. We'll have a grand old time one of these days.

'Night dear,

Bo

P.S. Green, who just went home, has been awarded the Distinguished Flying Cross. He earned it.

. . . .

Rogers with Major Russell's dog, Sito, at the Bellevue airfield.

[Bellevue]
September 21, 1918

Dear Lover

This *is* becoming a stronghold of inactivity and laziness—nothing to do but put wood on the fire and plenty of people to do that. Yesterday we almost did a show—were about to leave in the rain when the wing called up and said we were thru for the day.

Last night we had the usual movies. The wing has purchased a cinema outfit with some surplus mess funds, and we're to have pictures five nights a week instead of four. However, pictures aren't all the entertainment we have. Tomorrow and Monday the 51st Division concert party is putting on a show. The 51st is probably the best fighting unit in France. Composed of the best Scotch regiments (the Ladies from Hell) they put the wind up the Boche when they get under way.

Thursday night the Coldstream Guards Band will be here. Probably you know about the Guards Band and Major [Mackenzie] Rogan. They are to England what Sousa is to America. When they give concerts in London you pay fabulous prices for seats, out here we hear them for a franc.

We are inviting guests from other squadrons and various units nearby. I'm having Shapard and a couple of other fellows from his squadron over one night. . . .

All yours, dear,
Bo

. . . .

[Bellevue]
France,
September 24, 1918

Dearest Child

The last few days have been so exceedingly gay and giddy I'm afraid you've been neglected but not forgotten.

Sunday night there was a concert, only a short affair as the men had just come out of the line that morning and were rather tired. Last night they put on a revue affair and a short curtain raiser. Some of the people are very clever, they had good voices, snappy songs, an orchestra, and a girl who compared favorably with Dick Morgan. It seems he is a riveter from the Clyde, a driver in the army, and has the Military Medal. Really tho, Izzy, he made a lovely looking girl, very pretty brown eyes, fine features, nice complexion, quite graceful, and has a fine tenor voice. There were a couple of other fellows who made quite passable looking girls, too.

Tonight they put on a vaudeville program, and it made quite a hit. . . .

The weather decided to be decent for a day or so, and we have done two shows today. The one this morning wasn't very exciting, a few Huns about but all low down and not at all offensive. This afternoon there was plenty of excitement. Our bombers were followed home by about thirty Huns who played the dickens with them. We went down on the Huns and had a nasty scrap. I had the C.O.'s machine, as my own is getting a new engine. At the first burst I found the Vickers gun wouldn't work, so I cleared off and let the others worry. However we got off very nicely, lost no machines and got a Hun or so.

Tomorrow we have to go all the way to the objective with the bombers, a matter of nearly forty miles. That's quite a way when you figure it's farther than from Palo Alto to the city [San Francisco]. We're all praying for dud weather.

Good night, dear Isabelle.

Bo

. . . .

[Bellevue]
France,
September 27, 1918

Izzy Dear

For the last day or so we've been very busy, shows every day, shows every night, and plenty of excitement. Wednesday [September 25] we had two patrols, one in the morning and one in the evening, but neither very exciting. Yesterday we doped out a little plan to beguile the wily Hun, but there were no Huns about, so we came home rather fed up.

The wing washed us out in the afternoon, so took my machine with a new engine for a work out and flew up to see Bob Lytle who is about sixty miles away. I landed at an aerodrome in his district to make inquiries, and they told me that the place I was looking for was only a few miles away. Alas, the new engine wouldn't start, to be technical it wouldn't jump from the compensating jets to the main jets of the carburetor. All the fool thing would do was choke, pop, and sputter. I gave it up, sent for an engine man, and went in for tea. After the jets were jumped and the kite was running I figured it was best to keep it running, so I went home.

Last night we had the Coldstream Guards. In spite of the fact that Major McKenzie Rogan apologized for only having fifty pieces they made a lot of music.

The major is a charming old gentleman, eighty-two and looks about sixty, tours around France with his band, and proudly states that he has been in the army for fifty-two years. He was a great friend of King Edward's and has all sorts of decorations.

I always have had visions of the Guards in red uniforms and great bear skin shakos, as I believe they do wear in peace time, but last night they all wore commonplace khaki and look just like any other guardsmen. But such music!—I'm enclosing the program. Besides they played for encores Sousa's "Stars and Stripes Forever," "The Men of Harlish," and "Tipperary." They simply tear marches to pieces—plenty of brass and lots of drums and cymbals. At the end they played the three national anthems—French, American, and British—and everyone went away sort of subdued.

Somehow or other a band always gets under my skin. It would be terrifically easy to do all sorts of wild things to music like that. But fighting in an aeroplane is a silent and lonely sort of business, nothing to hear but your own engine and no one to talk to or be with.

Speaking of fights we had a lovely one this morning, and put a Boche or two out of working order. The war started again today and we were escorting bombers on an expedition to a Hun aerodrome. As usual wind and sun were against us. Over the objective seven Huns attacked the bombers, and we went down on the Huns. I dived on one who half rolled away when I started shooting. I pulled out in a left hand climbing turn and right in front of me was a Hun in a stall, nose up almost vertically and the machine scarcely moving. It was a lovely target so I gave it to him with both guns. He slipped out, then burst into flames.* It's a nasty sight, Izzy, even if it is a Hun, but you're sure you have them that way. About ten seconds later Lawson got another in flames. I went down on another Hun who was after the bombers, scared him away, and then got below everyone. The whole squadron was going around and around, more Huns came up, and being unable to get back into the scrap came home under the bombers.

Two other fellows got Huns out of control, and Lawson hit one with his undercarriage. It probably broke up in the air. Everyone got home so it wasn't a bad show.

However we have another this afternoon. I hope we don't see anything for I've had excitement enough for one day.

No more now, dear, for I must test my machine and the new carburetor

---

*Rogers's Combats in the Air report for September 27 claimed a Fokker biplane destroyed over Emerchicourt at 13,000 feet. The Sixty-second Squadron confirmed two enemy aircraft descending in flames during that engagement.

that was put on this morning. If the war keeps up we'll probably be fairly busy for a few days. But then the weather may turn bad again, who knows, who knows. Surely not your old lover.
Bo

. . . .

[Bellevue]
32 Squadron, France
September 29, 1918

Dear Child

Primarily there were two letters from you this morning.

You know, about the newspaper clipping mother sent you, I issued orders—yes! orders—some time ago that publicity should cease forthwith, but my brother-in-law [Ivan St. Johns] got hold of the wild tale and must have told it to some of his newspaper friends. Ain't it awful this "barely twenty" stuff. Once they had it "just nineteen" which is a trifle worse. I'm really getting to feel terribly old and worldly-wise tho, dear.

And as for Major Thompson there was a time when I might have been able to look him up on the Marne but not now.

This morning we had no scheduled show. Altho Sunday is *the* morning of all mornings to sleep late I crawled out at seven thirty, had breakfast alone, got the old kite wound up, and went forth for to strafe the wily Hun. It was an ideal morning for two-seaters to be over so I climbed up to 18,000 and started up the line. Right away over comes a two seater well above me and several miles ahead. I started to climb like the dickens. In the meantime he was busy taking pictures of a large part of Northern France. Finally he turned for the line again, apparently saw me, and turned away again to get a little more height. I was just about at the ceiling and staggering along when the Boche started for the line again. He was a couple of hundred feet above, and I started shooting as he was coming toward me hoping to get him to lose a little height. Then my Lewis gun jammed, busted beyond any fixing. The Hun passed by overhead, and I swung in under his tail and started shooting with the Vickers. The old boy turned then and the observer cut loose, but his shooting was very wide and didn't come anywhere near. So around we went, both of us popping away in great style. Then just when I was beginning to get in close the Vickers developed a bad crossfeed and I pulled out trying to fix it. It was incurable, and Mr. Hun sailed away for home. I came home nearly in tears for it's darned aggravating to have a chance like that, and then have your guns go flooie.

I've had the old wreck on the range for an hour and apparently the guns are alright, but you never can tell when they're going to let you down.

Yesterday afternoon I went down to call on Shap who has just come back from leave. He was to be found in a nice warm Hun hut that he and his roommate have made a few improvements on. His squadron is on an old German aerodrome in a desolate looking spot. He told me all about his pursuit by many Huns. It was funny to hear him tell it, but I guess he had a pretty bad time.

The day before [September 27] we went over on another show waiting for some bombers to come back. The bombers finally came with an escort of Bristol Fighters above them and dozens of Huns trailing along behind. They always remind me of a lot of kids following the calliope in a circus parade, first our bombers in a close formation and then a mob of Huns. These fellows were feeling pretty offensive and had a perfect right to as there were about forty of them, below, on our level and above. One little devil sailed along parallel to me about a thousand feet below until I got nervous and tumbled down for a shot or so. He did the usual half roll, and I suddenly discovered two more on my tail, and both of them apparently intent on doing me bodily harm. They were pouring streams of tracer in my general direction and suddenly the third one added his crack-crack-crack to the chorus so I came home in a hurry. They can't do you much harm if you're watching them, but their guns make a nasty noise and I always hunch myself up in the smallest possible space.

Yesterday we started a ping pong tournament, one franc entrance fee, all of which went to the winner. The major won and was awfully bucked. He says he needs the money to pay his bridge losses. We've taken him down a couple of times lately.

And now it's raining, the fire's roaring in our homemade stove, and I wish you and I were sitting in front of a big fire somewhere, but then I wish lots of things that can't come true now. They will someday.

That's all, Izzy dear, except an armful of love.

Bo

11. YPRES AND LYS *October 1918*

The front line crawled eastward while the RAF, frustrated by the weather, grasped at every break in the clouds to pound the retreating enemy columns. Canadian infantry pushed into the outskirts of Cambrai on October 9. East of Ypres, site of the bloody fighting of 1917, the British Second Army, despite rain and mud, crossed the Lys on September 19. The advance in Flanders secured the line from Ostend on the North Sea coast south through Lille and Douai.

Armistice rumors flew. Bulgaria had surrendered, leaving Austria-Hungary's southern flank vulnerable. In the waning days of October, the Allies were forcing the German armies into a narrow corridor in Belgium, where a double-track railroad between Namur and Liège was the main escape route across the Dutch frontier into Germany.

A break in the month's abysmal weather put the RAF into the skies to batter the rail junctions that fed this choke point. Near the end of the month, the Thirty-second Squadron relocated once again to a bumpy little field at Pronville, west of Cambrai. On October 30, an all-out effort by the Germans to prevent the day bombers from reaching their objective precipitated a mighty clash. Forty-one British planes were destroyed, with a loss of thirty-seven killed wounded or missing.

. . . .

[Bellevue]
32 Squadron,
October 1, 1918

Izzy Dear

October has come and with it a few things that might just as well have been left behind.

It's cold, very cold, and the oilcan stove is working overtime. We had a show this morning at seven, and I nearly hurled a death while emerging from my lovely warm bed into the bracing morning air. The old airships didn't seem to be very enthusiastic, either, for it was a dickens of a job to get them started. When it begins to get cold they take the water out of the radiators at night, put hot water in in the morning, use non-freezing oil in the guns, and take all sorts of precautionary measures against cold.

We did an offensive patrol and were to wait for some bombers to come home from a rather long distance raid. It certainly was frigid up above. In spite of all sorts of gloves three of my fingers have become permanently numb. The handsome Rogers physiognomy seems to hold out pretty well. About the only thing that happens is that my lovely nose gets disgustingly red. Poor old Tancock got pretty badly frozen this morning. A lot of the fellows have taken to whale oil and cold cream for protection.

We didn't have a lot of excitement—saw quite a few Huns but they must have been as cold as we were and didn't seem offensive. Just as we were coming home one fellow, a sleepy sort of a nut, got way below. We happened to look down just in time to see one Fokker shooting him up properly. We went down to the rescue and chased the Hun off. Then we tried to get him, but he was a stout man and knew just what he was about. Three of us chased him around in circles for ten minutes and never got one decent burst at him. Finally Zink got pretty close, so the Boche spun down and went home.

* * *

Signifying a lapse of time.

The bombers must have hit something this morning for much to every one's surprise the wing called up shortly after noon and said there would be no more war flying for the day.

As it was quite early I cranked up the balloon and went out to look for Bob Lytle. After a sixty mile run I came to the place where he should have been. It was a wicked looking aerodrome, large but sort of covered over with water from the last rain. Landing was accompanied with much splashing of mud and water all over me and the machine. But Lytle's no easy person to locate. He wasn't there. After much searching I found him. Of course there was a happy reunion and all that sort of thing.

Bob has been working on the present push with his own machine but under a British squadron. He proudly states—and it's probably true—that he is the first American trained pilot to fly over the line in an exclusively American made machine. Everything on his bus, guns and all, were made in those dear United States.

He's coming down here one of the days for a visit, as everyone is anxious to see the Liberty motor and what it can do.

I didn't get away until after six. By then I couldn't see more than a mile or so in any direction. However I set a compass course and after getting south a way was able to pick out plenty of familiar landmarks and came directly to the aerodrome. After a time out here one sort of gets the homing instinct.

About noon a nearby archie battery cut loose and we finally located the Hun miles up. It looked like an ordinary reconnaissance, but a little while after he'd gone a lot of little white pamphlets came floating down. What do you suppose they were? Peace propaganda. I wasn't lucky enough to get one—only a few fell here—but they contained the Austro-Hungarian peace note and a lot of bunk about humanity. At present most of the propaganda the British send over consists of well directed 112 pound bombs. There's no foolishness about them, and the Hun knows pretty well what they mean.

But things are really most encouraging, dear. The push up north is going splendidly—in fact things are good all along the line. The Bulgarian affair practically puts Turkey out, too, and should make a very deep impression in Austria. The end surely is in sight, but it's not here and I can't see how it can come before next spring at the earliest. In the meantime it gets colder—.

I'll keep an eye open for the 110th Infantry [Major Thompson's unit; letter of September 29]. They may be around here. I understand the Americans got in a bit of a hole here—pushed ahead very rapidly but failed to mop up properly and were cut off. I don't know how they have come out yet but they ought to be able to fight their way back.

I want to see you so badly that it hurts.

'Night, lover dear,

Bo

. . . .

[Bellevue]
32 Squadron,
October 2, 1918

Dearest Isabelle

Two more letters from you today.

You're getting to be quite popular singing all the time. Among many other things you *can* sing. And speaking of the "Star Spangled Banner" we all sang "God Save the King" when the Guards Band was here, but much to my horror I discovered too late that I didn't know it. Imagine one not knowing one's national anthem. Standing in the front row with Major Mackenzie Rogan leading the outfit, all I could do was throw back my head and open and close my mouth. Somehow or other it seemed as if he was looking directly at me all the time. Must have been my guilty conscience.

At present it's very wet and cold. We had a show this morning, but it was cloudy up high and all we saw was five Huns and a few bursts of archie. After coming home from an early show we always have breakfast again and clean up a bit. After that there's a bunch of official stuff that comes to the office, intelligence reports, the daily wireless press, the line, and a lot more that we look over. Then everyone hangs around for the mail. There really shouldn't have been much mail as Wednesday is usually a bad day but there was and I wasn't a bit disappointed.

Most of the afternoon I've been playing bridge with the Major, Zink, and Callender. Zink and I finally managed to get away with two francs. Then Zink and the major started playing poker for five francs a hand. You see, there are two terribly old and torn five franc notes in circulation around here and another five consisting of a couple of dirty one franc notes and the rest in copper and French twenty-five centime pieces. Everyone tries to get rid of them. Zink had the whole lot. As he lost all the poker hands the major now has them.

The war is lovely—our people in St. Quentin and things in good shape all along the line. Maybe the Boche is getting a bit worried.

This is all now, dear. I've a date with the barber—said barber having been borrowed from another squadron. Probably he's a blacksmith and only cuts hair as a diversion.

Wholesale love, Izzy dear,
Bo

. . . .

[Bellevue]
32 Squadron
October 5, 1918

Dearest

We haven't done many shows in the last three days—only one tame affair minus Huns and almost minus archie.

Yesterday we waited around most of the day for a show that never took place owing to a high wind and low clouds. About five I went down to visit Shapard who is still in the godforsaken area. Callender went along with me, and we perched on their horrible aerodrome without breaking anything.

As we were having a show [performance] last night, I asked Shap if he could come back with me, and his C.O. gave him permission to fly up.

"A" flight deputy leader Rogers, as seen from Alvin Callender's plane.

But my airship wouldn't start. Everyone took a try at swinging the prop, but something was radically wrong. Finally Callender went ahead, and we got hold of a sergeant who located the trouble and fixed it. By then it was getting dark, but we started anyway as we hadn't very far to go and there was no chance of going wrong.

About ten minutes later some heavy clouds darkened things up considerably, but I was following my landmarks easily and Shap was only a few yards behind. The exhausts began to pour out streams of blue fire and it became steadily darker, and I couldn't see any of my instruments. In trying to see them I turned off of my course, lost my landmarks, and didn't know where in the dickens the aerodrome was. Shap was still behind, following like a little dog.

I started firing lights—red ones and white ones and green ones. About ten seconds later lights started up from three separate aerodromes, and I didn't know which was ours. At one place they put out landing flares.

Finally, I saw a familiar road, shoved the old nose down, and headed for home in a hurry, Shapard still following. We landed in the dark, a very scary business, but both got away with it.

About the only excitement of the day was when some chap from another squadron nearly hurled a death just outside our field. He was taking off when his engine failed. As he was only a few feet up, there was nothing to do but go straight ahead and pray a bit. Unfortunately there were trees and a road ahead. When he tried to go between two of the trees he left his wings behind and landed in the road more or less on his head. The machine was a mess, but he wasn't badly hurt, only a broken nose or some such thing. The machine started to burn around the engine so we all rushed out with fire extinguishers and had a fine time—squirted them on the fire until it was out and then squirted them on each other.

There's a large cup of hot chocolate waiting here and after that, sleep. But sleep or no sleep I love you lots and lots, Izzy dear, and don't you forget it.

'Night,
Bo

. . . .

[Bellevue]
32 Squadron, France
October 7, 1918

Dear Child

We haven't done a show in days. It's been too bad for any kind of flying. We were to have done a rather long escort this morning, but ole Jupe Pluvius turned on the faucets up on Mount Amphibious and the party was off.

Tancock and I decided yesterday that our respective healths and good looks were in danger of a relapse from want of exercise, so we have started a course for physical improvement. Yesterday afternoon we dug up an old pair of gloves and boxed. We must have overdone it slightly, for we were both a little stiff today and took a long walk in the rain instead. All dressed up in field boots and rain coats we roamed over the hills and thru the mud in search of rabbits. Tanny had a Hun rifle, and I a revolver. We finally spied one sly old bunny who ran miles when we started shooting at him, and we plodded after. Finally he stopped, but as soon as we got anywhere near away he went again. We decided that hunting French rabbits was too

Lieutenant Monte Tancock, Rogers's deputy flight commander, November 1918.

strenuous, so amused ourselves by shooting at tin cans. Tin cans don't run when you shoot at them.

This evening after supper I yielded to temptation, got tangled up in a poker game, and am sadder, poorer, and wiser.

That's all, except that I'm getting darn well fed up with not having seen you for all these months and months.

Lots and lots of love, dear,
Bo

. . . .

[Bellevue]
32 Squadron,
October 12, 1918

Dearest

Can you imagine any worse catastrophe than Rogers with the last drop of green ink gone, the bottle dry, and nary a drop to be had in this part of France? Gosh awful state of affairs—what!

Speaking of the weather any self-respecting Oregon mist or San Francisco fog would blush crimson with shame were they able to see some of the meteorological phenomena produced here in the last day or so. This afternoon the fog was so thick you couldn't see a hundred yards. Now it rains steadily and probably will forever and ever. Of course there have been no patrols and consequently the bridge and poker games are working overtime.

As for excitement there hasn't been much except that a visitor from another squadron hurled a rather ugly death just in front of our hangars yesterday morning—got to stunting near the ground and spun into it.

Last night was our regular guest night and I had a couple of fellows from an American squadron over. We went to the movies first and very good ones they were—wild west stuff with Indians and barroom brawls, and bandits. . . .

Yours completely,
Bo

. . . .

[Bellevue]
32 Squadron,
October 13, 1918

Dearest Izzy

Just a short Sunday service for you see writing to you is my chief form of worship.

Of course the big topic today has been the peace negotiations, or it may be better to say the armistice talk. It's encouraging news, but I'm afraid it has created the wrong impression among a good many, especially among the men. They seem to think that the war's all over. All this premature talk and reports might very easily lead to serious results. It's no time now to let down and leave the Hun alone. However just as a gentle reminder that the scrap is still on our operation orders came thru the same as usual this evening. We're to escort the bombers on a raid to a Hun aerodrome a good way over the line. And of course we are to leave the ground at six thirty.

The chances are pretty fair that we will, too, as it cleared up nicely this afternoon and the moon was shining an hour or so ago. Perfectly wonderful moon, dear—would be for you and me but out here it may not be so nice, as bright moons are conducive to night bombing.

Tancock and I had a long session on the range this afternoon and shot lots of the King's good bullets at bottles and such. However Tanny didn't use any of the King's cartridges, but some of Kaiser Bill's. He has a Hun rifle and plenty of ammunition. One should get to be a very fair shot after a few months out here for there is always plenty of opportunity to practice with either a rifle or revolver. We used to have shotguns and clay pigeons but something has become of the guns.

Aside from that I've worked on souvenirs, written a few letters, played a couple of rubbers of bridge, and put a wax polish on the frame that your large picture is in. Not a very thrilling day.

Early patrols necessitate early rising, and one must sleep. I'm about to sleep, lover. 'Night.

Much, much love,
Bo

. . . .

[Bellevue]
32 Squadron,
October 17, 1918

Dearest Child

There hasn't been a letter from you for ten days which is about nine days too long.

We're winning the war. There's no doubt about it. This evening there was more good news from headquarters—Lille, Douai, and Ostend, they say, and things in good shape everywhere along the line. I'm particularly please[d] about Douai, for the archie around that town has always been too good for all practical purposes.

We haven't done much to help the good cause along for several days, as the weather has been impossible. This afternoon we tried an offensive patrol but couldn't get to the line for mist and low clouds. Monday we did two very long escorts with the bombers. While archie was plentiful and accurate there wasn't a Hun in the sky. Doing nothing surely gets tiresome and everyone is itching for shows and a bit of action.

This afternoon Tanny and I went for a long walk over to an American squadron and didn't get back until nearly dinner time. A good walk now and then puts an edge on the appetite and fills you full of pep. It gets dark

Lieutenants F. L. "Buddy" Hale, Monte Tancock, and John Trusler. Hale, an American, was an ace; Trusler, the squadron's oldest pilot.

quite early now, and a glorious moon was up by the time we got back. At least it might have been a glorious moon, but the time was wrong and the place wasn't right and there was no girl at all, and that's really the most important of all.

    A chap named [John] Trusler who was the oldest man in the squadron is going home tomorrow to instruct or, if he can make it, go home to Canada on leave. These departures of people for home are always the occasion of a party. This evening has been no exception, quite a large party. Fact is, it's past midnight now, and we only keyed down a few minutes ago. Pretty wicked staying up past midnight don't you think.

I'm sending you a paper knife made from a little of everything and pray that you will overlook the fact that the handle is lopsided. Bein' as how it was the first one I attempted it's far from perfect. The blade is made from a wire from my airship, the handle is pretty much everything—washers, brass, copper, aluminum, the white is from little celluloid signboards which were in my bus and quite unnecessary, and the black stuff is part of a phonograph record.

Some day I'll make a decent one. And speaking of souvenirs I've some silk from a parachute flare which you shall have. I can't imagine what it might be good for, but some of the pieces appear to be large enough for handkerchiefs. However, when these curiosities are carried to one's wearing apparel it's probably going a bit too far.

Our move is beginning to materialize and will happen one of these fine days. The new aerodrome is well forward and not too good so they say, but I've seen some pretty bad ones out here and refuse to worry over these small details. But my stove, my lovely stove, must that be left behind! I pray not.

It won't be long after this letter arrives until you'll be thinking of Thanksgiving and even Christmas. Time is an awful nuisance, isn't it, dear, and yet it's something of a barrier.

Lots and lots of love, dear,
Bo

. . . .

[Bellevue]
32 Squadron,
October 18, 1918

Dearest Izzy

My ship came in this morning—but perhaps ship is hardly the word. Twelve letters came nearer to being a fleet. Two were bills which don't count, one from Dave Smith, one from mother, two from fellows out here, one from my sister, one from a chap you don't know and wouldn't know if I were to tell you his name, and four were from a girl in Oregon. Count 'em—twelve!

Having spent many happy minutes reading them I shall spend many unhappy hours answering them—all except the four. I don't dislike writing to you, dear.

Strangely enough every letter had been opened by the censor. In all the time I've been out here I don't believe more than five or six have been opened. But inasmuch as nothing had been censored I can't object. And the censor must have had a lot of good reading. I hope it was no lady censor that read the one from Smith.

Shapard ground out a short note full of woe. He is pretty well up front, and has been doing a good deal of low strafing. His latest adventure consisted of being shot in a portion of his anatomy where it's best not to be shot. Fortunately the bullet only ripped away some very essential portions of his clothing and didn't even scratch him. The boy is lucky. He earnestly requested me to prolong the war as long as possible, as Cox had just sent him a statement saying he was ten pounds overdrawn.

And speaking of the war we had another lovely official report this morning—Ostend and Lille taken and our cavalry near Bruges. The Hun will have to abandon the coast in a hurry. That means no more air raids on England and a much less effective submarine warfare. Still it makes very little difference to us. The farther the line goes back the farther we go over on shows. They seem to find any amount of targets thirty miles over. Too far.

Our move should take place almost any day, altho we maybe here for some time.

I'm trying to get rid of some of this laziness. Just to prove my sincerity I got up at seven this morning and ran a couple of miles before breakfast with the C.O. and a couple of other fellows. Being attired as scantily as possible without shocking the French civilian population it was rather chilly work but it surely makes the blood do more than ooze thru the veins. Also it made my cheeks a delightful pink and my nose a disgusting red.

It's still too dud for any sort of a long show, no clouds but a heavy haze that means no visibility at all when you're in the air.

I let my old airship full out for the first time yesterday afternoon and it's a lovely machine. Two miles a minute she does, and that's fast enough for all practical purposes. Also it's light as a feather on the controls and will dive up to any old speed. . . .

Completely yours, lover,
Bo

. . . .

[Bellevue]
32 Squadron,
October 19, 1918

Lover

The weather is getting on my nerves. It continues to be too dud for all flying even to playing around the aerodrome or making calls. Most depressing weather I ever saw, too, not cold or not warm or not cloudy—just plain still and thick. However it's started to rain a bit this evening and may change a little.

The drab day has had two bright spots consisting of two letters from you. . . .

Had a letter from George Crary who you probably don't know but who is one of the brethren in Kappa Sigma. He's just transferred to aviation. He said that his regiment, made up of Oregon, Washington, and California men, had a pretty bad time of it in the line but did very well. The fellows I knew in it came thru alright.

Good night dearest child,
Bo

. . . .

[Bellevue]
32 Squadron,
October 21, 1918

Dearest Child

There was a lovely long letter from your mother yesterday. She writes wonderful letters, Id, and they're quite easy to answer. We get pretty clubby in our correspondence with each other. . . .

Here begins my heartrending tale of woe, a story calculated to soften the hardest heart, to bring tears to eyes that have been dry for twenty years:

This morning some enthusiast suggested football (soccer) as a mild form of amusement and exercise. All regular players, we really have a very fine team which plays in the wing league, were barred. A & B flights were to play C flight and headquarters which includes everyone in the transport, orderly room, carpenter and machine shops, the armoury etc. The teams were about half officers and half men, and it was a wicked battle. I played a forward and nearly died before it was all over, but enjoyed every minute of it. Unfortunately we lost, 2 to 1. Now as I sit by my fire I feel a strange stiffness creeping into my limbs, the muscles ache, the joints creak. Also I have three large bumps on my right shin and two skinned knees. I fear getting up in the morning will be a painful process.

The remainder of the day has been spent between a walk before breakfast, don't be alarmed, we have breakfast until ten, a hair cut, and making a model of a Fokker Biplane. The C.O. wants models of all Hun machines so that new pilots may know what they are like. Two of us are making a collection of scale models. It's a bit of a job, but very interesting. Also there is any amount of material to make them out of and a fine carpenter shop. Any sort of work that will keep time flying by is welcome these days. It doesn't show any signs of clearing.

Sometimes I wish I didn't love you so much, dear—then maybe I wouldn't think about you so much.

Wholesale love,

Bo

. . . .

[Bellevue]
32 Squadron,
October 23, 1918

Dear Id

As usual it is dud, altho not as bad as it has been. It may even clear up enough for us to do a show later. The ground haze is pretty bad, but the sun is trying to struggle thru and may succeed.

The move looms nearer and nearer, in fact most of the heavy stuff is packed and ready to go. We're only waiting for a decent location.

Yesterday Tanny and I went for a long walk and did a bit of shooting at some grouse and rabbits. Trying to hit them with a Hun rifle and a revolver is rather hopeless work. Also one has to be careful as the landscape is liberally scattered with French cows and horses and farmers and women and small boys. Every shot brings a shout of protest from a chorus of angry voices. The fields about here are very pretty now and the little patches of wood, also. These French people work continually on their land from dawn to dark, men, women, and children. Yesterday we saw girls milking the cows and old, old women digging sugar beets, and old men and young boys doing the plowing. There are no young men running about.

You must know about the cunning little mice that nearly ruined all my clothes—I had a large woolen scarf that my grandmother knitted packed away in the bottom of my bomb box bureau. A day or so ago I noticed a cute little mouse emerge from said bureau. Being a suspicious soul I at once removed everything and found that the dear little mouse had eaten a nest thru three layers of this scarf and that ole Doc Stork had deposited three small mice there. The scarf is no more, nor are the mice. Rats and mice are a terrible pest out here and will eat anything and everything.

Yours,
Bo

. . . .

[Pronville]
32 Squadron,
October 28, 1918

Dearest Child

Gosh awful place, this. We came down yesterday afternoon and have been busy ever since, getting settled and things fit to live in. It's not what you might call a lovely spot, in fact it is one of the most unlovely places I ever hope to see. Only a few hundred yards away is part of an old Hun trench system that was very strong and very strenuously fought for. As a result the whole place is pounded flat and most desolate looking.

This afternoon we went on a little exploration thru trenches and into dugouts and old gun emplacements. There are some wonderful dugouts — thirty or forty feet deep many of them, strongly lined with timber, and very roomy. There was also a good deal of concrete used in some of them and, of course, the customary sandbags.

There is all sorts of material about, live shells and dud shells, bombs ranging from dud Boche aerial bombs to live hand grenades, old rifles, helmets, shell cases, everything you can imagine.

On a nearby hill are two villages, both badly knocked about. In one of them seven Scotchmen held up a heavy Hun counter attack for two days. It was considered one of the most heroic exploits of the war, as they were entirely cut off and pretty short on rations and ammunition. One of them got the Victoria Cross and the others received lesser decorations.

The neighborhood isn't exactly safe yet. Mines go off most unexpectedly where they have no business to, and there were a lot of booby traps left by the Hun. You pick up an innocent looking rifle and are blown sky high by a bomb.

When the wind is right—or wrong—it carries the none too pleasant odor of horses deceased long ago.

The aerodrome? No. Just because hangars have been placed around a bumpy little field and a landing "T" placed in the center is no reason it should be called an aerodrome. You have to figure things out just right to

Callender inspects a dud German aerial bomb, Pronville airfield, October 1918.

get into it. Everything is under canvas. We have bell tents which are put over dugouts, if you get the idea from this beautiful picture [hand-drawn sketch].  We are two feet in the ground with a tent overhead. They are very roomy and easy to keep warm so might be worse. . . .

This morning we did a show with the bombers and ran into beaucoup de Boche—too many, in fact. They went for the bombers, we went for them, and they went home. On the way back we met more and had a tiff or so.

Last night after supper the C.O. took me for a little stroll and said that he had recommended me for a captaincy and that the wing had passed it. I'll let you know when it comes thru officially, and you can address letters to Captain B R etc.

Of course recommendation means waiting until there is a vacancy somewhere. It may be here, it may be in some other squadron out here, or it may mean going to England for a flight. I surely hope it will be here for there isn't a better C.O. or a better squadron in France.

Lots of delayed sympathy for the night rushing was over and more than lots of love.

Yours,
Bo

. . . .

[Pronville]
32 Squadron,
October 30, 1918

Dearest Lover

. . . We had a bad day today. Got into a terrible mess this morning bringing the bombers home. They all got back, but we got properly into the soup driving Huns away from them and had to stay and fight more than twice our number. There were Huns everywhere, above, below, and on both sides. Every time you'd look around there would be more of them coming up. I had two of 'em worry me almost to tears, one above and one below. The one above kept coming down and taking the odd shot, and the one below got rid of some spare ammunition, too. I couldn't go down on the bottom one for fear of the top one and there were seven more coming up behind. Everyone else had their hands full. We have a man or so missing and another chap down in a C.C.S. full of holes.

We did another show this afternoon, but didn't see anything, thank heaven. One large party a day is plenty.

It's getting cold, very cold in fact, and the minute the fire is out our old tent is a refrigerator. As it's bound to get cold at night anyway, we open everything up and have plenty of fresh air. I'm a bit of a fresh air fiend anyway, windows open, sleeping out of doors and all that sort of thing. We aren't doing much fixing up, for we may make a change for the better very shortly.

No more bunk tonight, Izzy dear. I'm getting fed up with this war. Wish the damn thing would stop.

Much love,
Bo

P.S.   C.C.S. = Casualty Clearing Station

12. ARMISTICE *November 1918*

An influenza pandemic ravaged the globe, killing by conservative estimates 27 million people. Imperial Germany teetered on the brink of revolution. All along the Western Front, the Allies pursued the German rear guard. When weather permitted, the RAF bombed and strafed retreating troop columns, railroad stations, and airfields. On November 2, the Thirty-second Squadron, reduced to nine pilots and seven aircraft, moved to La Brayelle, an airfield near Douai. On November 7, New York headlines announced an armistice, which proved to be false.

On the evening of Sunday, November 10, Canadian infantry cleared the last pockets of resistance out of Mons, where the British Expeditionary Force had fought its first battle of the war in August 1914. Kaiser Wilhelm fled to Holland, and the German representatives in the railway car in the forest of Compiègne prepared to sign an armistice. At the stroke of the eleventh hour on the eleventh day of the eleventh month of 1918, the Great War became history.

. . . .

[La Brayelle]
32 Squadron,
November 4, 1918

Dearest

What a mess I've made of letters the last few days. Not a single one have I written. There hasn't been one from you in ages, but I know it's not because you haven't written. Fact is, I've never felt less like writing than I have the last few days, and on top of that there has been very little time.

This is the last place I ever figured on living, a German hospital, but we're very comfortably settled having made a change for the better, have a lovely place, and may really stay here for a while.

Our present quarters are in an old French chateau that has been used as a Hun hospital since 1914. The place was quite clean when we came in and pretty well intact, only a couple of stray shell holes in one wing. Downstairs there is a large mess hall, formerly an operating room, a smoking room, great hall, and a very nice kitchen and pantry. The interior is all carved oak paneling, but has been enameled white. The whole interior is the same. The rooms have large open fire places and big mirrors over the mantles. Upstairs there is plenty of room for everyone. Three of us have a corner room, intact except for one window and a cluster of bullet holes in the wall. The window has been covered with oiled silk and a picture hung over the holes, so you'd never be able to tell there was a war. We have a large stove, plenty of rugs and furniture, and mirrors galore, a full length affair against one wall, a large carved oak one over the mantle, and numerous other smaller ones scattered here and there.

The grounds must have been very pretty before the war, all lawns and trees and gardens and marble flower bowls and summer houses and a high brick wall around it all. However, they are badly cut up by dugouts and a light railway track and much rubbish. Fortunately the rubbish consists mostly of wood, which will come to an ignoble end in the numerous stoves.

Out by the stables we found a green house full of strawberry vines and a neat little garden full of lettuce, beets, and carrots thotfully planted by the Boche. I tell yuh! — fine spot.

The aerodrome is one of the best I've ever seen — turf like a lawn and acres of it.* It's a prewar affair and has a history which the censor would never stand for. Altogether its not to be compared with the last place. I hope they don't move us again soon.

We've been having a rotten time of it, another awful scrap a couple of days ago. We were lucky to get back at all. A couple didn't. I managed to get another Hun.† I'm pretty sure he was done for, but then five more chased me all over the shop. Too many! Today we did two shows, both of them very dull and cold. It was a wonderful clear day, the best in a long time, but intensely cold up above.

Yesterday two of us went down to the aircraft depot for new machines. We rode down in the C.O.'s car and came back in a wicked rain storm, clouds so low that we practically had to hop trees all the way home. We

---

*A former tenant of the airfield at La Brayelle was the Red Baron of Germany, Manfred von Richthofen.

†Rogers's Combats in the Air report for November 1 claimed a Fokker biplane driven down out of control six miles east of Valenciennes after he went down vertically on it, firing a burst of of fifty rounds at point-blank range.

Manfred von Richthofen's cottage at La Brayelle airfield.

came very nearly staying there all night, but it's a most inhospitable place, and after thinking about our nice soft little beds and our large fires we took a chance.

In about two more days, if things go right, you may drop the Lieutenant and try Captain for a change. Poor old Callender got into a bad hole a few days ago and died of wounds in a Canadian field hospital. I've been put in for his flight and should know definitely in a day or so. If it comes thru, it means staying out here for another two or three months and doing some real work. Otherwise I could have gone home or to England to instruct. However all these small advantages have their disadvantages. See Emerson: "Compensation."

I do miss your letters and get all depressed and disagreeable when something happens to the mail.

This isn't much of a letter, but the next ones will be better. I've had too much enthusiasm knocked out of me the last few days to be entertaining.

Lots and lots of love from
Bo

. . . .
                                                                      [La Brayelle]
                                                                       32 Squadron,
                                                                   November 7, 1918

Lover Dear

The sweater worries me. It fits perfectly, the color couldn't be better, and there is, as usual, that tantalizing fragrance about it that starts me thinking of all sorts of things.

Had a letter from Dutch Koerner today, the first in several weeks. He has been in the line, went over the top five times, and is still in one piece. He said his division, mostly Oregon troops, had done very well altho it was hard going. He's near here somewhere now, so I'm going to try to find him. Deke Gard, who you probably know of, was killed altho Dutch said all the rest of the Stanford fellows came thru nicely as far as he knew. . . .

The weather remains dud. This morning the C.O. and I got very ambitious and filled up holes in the aerodrome. We're a bit short handed now, as there has been an epidemic of "flu" which is about over, thank goodness. They've surely had a lot of it out here and in England, too. Guess it's everywhere for the states have had their share.

And this concludes the day's performance, except for the incidental fact that I've not ceased loving you one small bit.

'Night, lover,
Bo

. . . .
                                                                       [La Brayelle]
                                                                       32 Squadron,
                                                                  November 10, 1918

Dearest

Tonight is a night of much excitement—arguments, predictions, prognostications, auguries, prophecies, and bets, all as to whether the Hun will say yes or no tomorrow. My poor old head is all befuddled so I've decided to argue no more and to wait until tomorrow. What he says he'll say and I'm tired of this wild and senseless debate. However if he doesn't say "yes" I'll be disappointed, and so will a few million others.

We've been working reasonably hard the last two days—two shows yesterday and two more today. They were all uneventful, no Huns about altho quite a good bit of archie.

This afternoon's show I led myself, felt quite important with four of my own men and two other flights following. The official captaincy seems to have become lost in the shuffle, but ought to be thru in a few days. I'm posted as temporarily in command of the flight.

There aren't a bad lot of pilots in it. Tancock, who I've spoken of before, is my deputy leader, there's one wild Canadian boy who is full out for trouble all the time, one wise old reliable Canadian who is always on the job and absolutely dependable, two South Africans who have just joined the squadron, lean, ugly looking men who are both regular persons, good pilots, and very keen about the work, and a Scotchman who is the least good of the lot. Only two of the lot have had much experience, but they'll all learn quickly enough.

No leave for another two weeks—I was due to go today, but will have to stop over and do a little honest labor....

I must write to Simmie and Green and Allan Crary and tell them I'm not coming on leave, so—.

Much, much love,
Bo

. . . .

[La Brayelle]
32 Squadron,
November 11, 1918

Miss Isabelle G. Young
KAT House
Stanford University
California
U.S.A.

Well Dear!

It's all over. I surely thank God it is and that I'm here to see the finish. There have been more times than one that I've thot it would be otherwise.

Last night there were vague rumors that the Hun had quit, but operation orders came thru as usual. This morning we were standing by for a show, everyone sitting around the breakfast table sort of waiting for news. It finally came thru from the wing "Hostilities are to cease at eleven o'clock. No war flying is to be done after the receipt of this message."

Somehow or other it didn't seem possible it was all over, that we were thru with crossing the line and worrying about Huns and archie and seeing good fellows getting bumped off.

We went down to the aerodrome, had all the machines put away, and told the men to do as they pleased. About an hour later we found most of them in the canteen drinking horrible French beer and singing at the top of their voices. So we bought beer for them and drank it with them. I hate beer before lunch.

Tonight a big party is on the boards. I'm afraid it will be wild and wooly. They don't have days like this very often and one needs [to] celebrate. We'll probably end up in the sergeants mess. That's where all the large parties end.

The immediate future is a mystery. Nobody seems to know just what will happen, whether we'll stay here or move to the frontier or what. As for demobilization, that's also a matter to worry about. It must be a couple of months or more before people start getting home. It's an enormous task and the poor devils who have been out here for four long years should have some preference. We'll probably begin to learn something definite in a week or so.

Izzy dear, I guess there have been a lot of things that I've never told you about. It's all been pretty awful. When I think of all the fellows who aren't going home, I wonder what right I've had to live. I know that surely there has been divine protection. And the way some of them had to go. Poor old Callender, as square and decent a man as ever lived, going only a week or so ago and Bill Leaf, who was killed only last week. Even yesterday one of the bombers we were escorting went down thru a direct hit by archie, almost a million to one chance.

People can prate until the judgement day about war being the salvation of nations, the one thing that can keep them from decay, but I know that it will never be worth the sacrifice. It's all wrong.

Whenever I think about getting home, well, words have their limitations. I know of no words that are expressive enough for the occasion. I certainly hope they get a move on. Steerage will do very nicely if there's nothing better. However, excessive optimism is a bad thing. You mustn't forget it will be some time yet, dear.

It's been a wonderful experience for the fellows who have come thru with a whole skin and somehow or other I imagine we'll sort of feel ourselves to be a superior race, altho almost everyone should be on an equal footing. I say "almost" because there are a choice few who have lost caste entirely. After all I suppose what you did in the war won't buy any bacon for breakfast a year from now.

Thank goodness I've seen a small bit of the world and taken a chance. At least I'll have a clear conscience. And, of course, in the days to come it will be nice when someone asks "Papa, what did you do in the great war?" But that's taking a great deal for granted isn't it, lover.

Remember dear child that from now on nothing counts but getting back to you. Thank heaven the chance that I might not is gone.

Yours completely,
Bo

. . . .

[La Brayelle]
32 Squadron,
November 13, 1918

Dearest Izzy

I've always maintained that with the elements of danger removed this would be a very nice war, and so it is. Strangely enough the weather has been perfect, clear, sunny days with the air just nicely brisk. As for this evening, twas on a night like this Leander swam the Hellespont.

This morning I sent some of my new people up to fly and fooled around the aerodrome generally. Then a couple of fellows from another squadron flew in for lunch and we had quite a small party—six or seven guests.

This afternoon three of us went for a long ride in the C.O.'s car and ended up in a very large city [Mons] recently evacuated by the Hun. It's a real city, full of civilians, and scarcely knocked about at all. We had tea and a look at the place. All of the civilians seem to be dressed in their Sunday best and are holding a sort of prolonged celebration. We talked with one little urchin about the Boche. He said that while they held the town, ever since fourteen, he hadn't had overly much to eat.

During the drive we went thru many villages that were in fair condition but all painted up with Hun signs and notices. Also there were several Hun aerodromes, some occupied by our squadrons and others vacant. We passed one that was raided in a big show Shapard had taken part in. We could see some of the results, hangars burned and bomb holes everywhere.

All of the roads are in very fair shape except for the destroyed bridges, and there's none of the awful desolation that is to be found in other areas.

The party celebrating peace was a huge success. I'm thankful we don't have peace very often. We started with the king's health, went right down the line forgetting nobody. Everyone managed to make some sort of speech. The parade ended up in the sergeants mess. Thank goodness such celebrations are rare.

Our future is obscure. Don't know what we'll do or when, but we'll very likely stay here for a time.

The captaincy is still tied up in red tape, but I live in hope and expectation. . . .

'Night, lover dear,
Bo

. . . .

[La Brayelle]
32 Squadron,
November 15, 1918

Dearest Lover

. . . This afternoon three of us went forward in the C.O.'s car with a cross for Callender's grave. Whenever we know where a pilot is buried we have a cross made from a four bladed propeller and a carved oak plate put in the center. You may be able to get an idea from the lovely sketch. Also we plant some sort of vines on the grave.

Today we had the map location only in a vague sort of way. After a long ride over terrible roads we came to Valenciennes where we made a few inquiries and then went out looking for the grave. By the time we got there it was dark. Altho we searched for half an hour we couldn't find it. Apparently there had been a field ambulance there which had moved forward later. There was no sign of his machine altho we rather expected to find it nearby.

The future is still obscure. We probably shall not move forward but back near the aerodrome we left three or four weeks ago. We should be able to settle down for the winter and a good thing, too, as there is plenty of ice every morning now. It shouldn't be long before we have snow.

It will all be monotonous and dull, for there isn't much to the army when there's no fighting going on. Not that I'm keen on fighting, but you feel as if you were of some use then, while now I know of one or two places I'd much rather be.

No matter, I would join the army. The staff captain from the wing was around today and told me to be patient, as my captaincy should be thru any day now.

There's the packing to be done so good night, dear, and much, much love.

Yours,
Bo

. . . .

[Izel-le-Hameau]*
32 Squadron,
November 17, 1918

My Dear

I take my pen in hand etc. etc. to inform you that winter is here in spite of the fact that the old timers scoff and say "Wait until we get a little cold weather." Yesterday, today, and probably forever, the ground has been frozen hard and all the water is ice.

Yesterday we came here and found it to be quite a decent spot, nice aerodrome, permanent hangars, plenty of very good huts for everyone, and many farms where we can get fresh food. Just after we landed two or three of us went down into a nearby village and managed to get a very nice lunch omelette, fried potatoes, and such. But we digress.

Two of us have a whole hut to ourselves and are getting it in very comfortable shape. We're dividing it into two rooms, a bedroom and a sitting room. We brought all sorts of furniture from the last place and should have quite a shack before we're finished. Most of the day has been spent in getting the place weather proof, putting in new windows of oiled linen, stuffing up all the cracks, patching a hole or so in the roof, and setting up the stove. All of our furniture hasn't arrived yet, so we're still somewhat unsettled.

They are starting to "hot air." In the future we are to have parades every morning at *eight* with flight commanders in charge of their flights, inspections, practice flying, organized sports and recreations for everyone.

It will be funny for awhile, as nobody knows much about drill. And it will be tragic, too, getting up at seven every morning, getting properly dressed and shaved, and shivering in the cold. Heretofore when there has been no early show, everyone has rolled in to breakfast about ten dressed in pajamas, a flying coat, and flying boots. The men, too, have never had many parades. They've all had their work to do and been busy with their jobs. While there is plenty of discipline, it's never been of the parade sort.

The army in peace time seems to be a much worse place than the army in wartime.

The mess president, just returned from Paris where he has been buying stuff for Christmas, said the place was pretty wild, and there was some

---

*Izel-le-Hameau, the new quarters of the Thirty-second Squadron, was located about eight miles east of Saint-Pol and about the same distance northwest of Arras. In May, Bogart had joined the squadron at Beauvois near Saint-Pol. He had come full circle.

big celebration on for today. They say London was pretty spry the night the armistice was declared. I'd love to have been in San Francisco.

The future is just as obscure as ever. Nobody seems to know what is to become of us. If they'd only forget about these early morning parades all would be lovely.

Tomorrow or the next day I'll tell you all about the man who escaped from Germany. Until then—,

Wholesale love, lover dear,

Bo

. . . .

[Izel-le-Hameau]
32 Squadron,
November 19, 1918

Dearest Isabelle

. . . Last night we had a party for one of the squadrons that has moved here, nice dinner, with oysters, soup, fish, roast lamb, des[s]ert, nuts, etc. etc.

I'm beginning to feel positively ancient and noticed it more than ever last night. I used to know several fellows in this particular squadron, but all of them have either been bumped off or gone home. The present lot have come out quite recently. It's the same in our squadron. Aside from the C.O. and the equipment officer McBean, my roommate [Bruce Lawson] and myself are the oldest men. Only five other flying officers were out here in August. The recording officer is fairly new, also. Nearly half the flying officers haven't been over the line more than two or three times. When you've been thru the mill with fellows, been in all sorts of tight places where you've helped them and they've helped you, there is a mutual feeling that isn't possible with fellows who have come out just about as the war ended. They're all pretty decent chaps, but it would have been wonderful if all the old bunch were here to take part in the good times. They all earned the right to be in at the finish and it was rotten luck for them to have gone west so near the end.

Tomorrow I planned on going up to look for Callender's grave but there is a football match on with one of the other squadrons so I shall go the next day instead.

Oh yes! The man who came back from Germany.

We had an American attached to the squadron [Lieutenant J. O. Donaldson] who went missing in a show about three months ago. I probably told you about the show. It was on a windy day, far over the line. We

Lieutenant J. O. Donaldson, whose escape from Germany is described in Rogers's letter of November 19, 1918.

figured the two fellows who were missing had run short of petrol. Anyway, we had a letter from one of them a week or so ago saying he was in England and giving a few details of his escape. A couple of days ago he breezed in for a short visit.

He had landed in Hunland after a scrap with three Huns, one of which he shot down in flames. The night he was captured, he and another chap, who I knew quite well, escaped and started for the line. They managed to get nearly thru when a Boche wiring party caught them and brought them back. As punishment they were put in close confinement on bread and water for two weeks. Before the two weeks was up they escaped. They tried to steal a Hun two-seater from an aerodrome at night but were caught and only got away after a fight with the guard. Then they headed for Holland traveling by night and resting in the daytime. They finally reached the border and were three days getting thru the electric wire.

I never liked this chap personally and still don't, but it was surely a stout show and one that deserves a lot of credit.

The light has blinked twice which means five minutes more so good night, lover dear.

Yours,
Bo

. . . .

[Izel-le-Hameau]
32 Squadron,
November 20, 1918

Dear Child

After I'd played football this afternoon and cleaned up a bit and was sitting by the fire, I got to thinking how awfully long it has been since I've seen you, dear. Ages and ages it has been, and I'm fed up with it. If they'd demobilize the whole flying corps tomorrow I'd swim home if there was no other way.

The future is going to be pretty wonderful for us, Izzy, and I'm impatient for it to begin. There's still some of the past ahead, and the future officially will begin only when I get back to you.

I've been wondering just what I'm going to do and have been doing a lot of wild thinking. There's bound to be a lot of big opportunities in flying, not this crazy exhibition stuff, but good straight commercial flying. I've written to Andy Smith and expounded a few of my pet theories, so we may be able to get together on something.

Callender's crash near Valenciennes.

After all there's not much to worry about. With you to work for it can't be possible for things to go any way but the right way.

'Night

Bo

. . . .

32 Squadron,
November 23, 1918

My Dearest

In spite of the fact that there's no war, there seems to be plenty to keep one busy. Thursday three of us went up to find Callender's grave and had very little trouble. His machine was crashed not far away, and we got mute testimony of what must have happened. It was badly crashed in a small field and it's probable the poor chap died in the air, or at least was unconscious. The machine had been simply shot to pieces, holes every where and some of them from explosive bullets. As far as we could see only one shot had hit him but as it must have entered his lungs it would have been enough. The grave was beside a little farmhouse and not marked at all. We put up the cross and then sodded the top and built a little border of brick around the edges. It was a solemn party that came home.

Yesterday morning I had my flight out for a little formation practice.

Rogers with the cross for Callender's grave.

We tootled all over the country and had a look at most of the old aerodromes to see if they were occupied or not. It's great fun diving in formation and zooming up again. The machines are much nicer without a war load of ammunition, bomb racks, and such.

Last night the Major, Veitch (B flight commander), and myself were invited out to dinner at one of the other squadrons. After a very decent meal we went to the movies.

Speaking of Veitch, he's a most regular person. While he belongs to an Indian regiment and claims to be an Englishman, he's spent most of his life in no other spot than Puerto Limon, Costa Rica [the setting of *Pirate for a Day*]. Perhaps you've heard of the place before. Furthermore his father owns a large banana plantation. He doesn't know Pedro Diablo and has never heard his father mention making gunpowder from banana skins. Also he says that anyone who has ever been in Puerto Limon might use it as the setting for a tragedy or a melodrama, but never for a musical comedy.

Today has been more than full. The sun melted all the frost and ice except in the shady places; it has been warm and clear; and the air, balmy and cool, is worth tons of medicine. It might have been May in California or Oregon instead of the end of autumn. About eleven I pushed off for a little joy ride all alone. Incidentally I wanted to locate Bob Lytle's aerodrome and drop a note for him. I went up to the coast, then turned north towards Boulogne. From 7000 feet the chalk cliffs at Dover were quite distinct. Later I could see the whole Sussex coast line. It was great just floating along above everything, the old motor never missing a shot, and no Huns or archie to worry about. I finally located an aerodrome that seemed to be the one I was looking for. After diving down, I circled around a couple of times, dropped the message bag, and headed for home. After thinking it over I'm convinced it was the wrong aerodrome, but what does it matter, I'll go up again tomorrow.

I've never enjoyed flying more than I did this morning. It's a wicked old game. Once it gets hold you're gone, never will be able to get away from it. If you'd ever flown, you'd understand the feeling, dear. There's nothing else like it and there never will be. As for the element of danger, after a time it becomes as easy as walking. Surely it's easier than driving a car. I never give a thot to landing or taking off any more. Once you get one of the old kites in the air, you just let go, and they do the rest. One of these days you and I, and lots of others as well, will be flying all over the place and thinking nothing of it. That day isn't very far distant, either.

After lunch the Major, Veitch, and I took a shot gun and two revolvers and went forth for to strafe the wily rabbit and the even more wily partridge. We wandered for miles thru muddy fields and wet woods and came home with nothing but three large appetites. Ordinary walking is exercise, but walking with several pounds of mud clinging to each foot is real work.

We have a large parade in the morning, inspection of the men and their quarters and a lot of hot air. It means getting dressed up like a horse, having breakfast before nine, and then trying to appear severe and dignified when you really want to laugh. Somehow or other it always amuses me to look men over from head to foot, to see if they've shaved and polished their buttons. . . .

Good night, dear, and lots and lots of love from
Bo
P.S. Captain Bogart Rogers is my new address.

. . . .

[Izel-le-Hameau]
November 24, 1918

We played a team made up of N.C.O.s this afternoon and they beat us by one point, which wasn't bad. The field was muddy, and by the time we finished everyone was pretty well coated. It's good fun tho and helps pass the time away.

The C.O. has donated a cup for an interflight football series. There's much excitement over it, especially among the men. I've several good players in my flight, and we ought to come close to winning it. There are to be five teams, one from each flight, one from headquarters flight, and one from the remainder which includes transport, office force, and the rest. We'll probably be a month playing all the games off.

. . . .

[Izel-le-Hameau]
32 Squadron,
November 30, 1918

Dear Child

... The latest I've heard is that Grubby Clover was killed out here in a crash. Thank heaven all this war flying is over. I'm caution personified these days, never think of doing a loop or a roll unless I'm thousands of feet up and not many of them then.

Guess I told you we had a bad time out here with flu. About a month ago the squadron was in a bad way, over thirty men in hospital altho only one died. There have been 32,000 deaths in England in six weeks, so you see we know what it's like.

The censorship rules have been eased up somewhat and we are permitted to give our location. Izel-le-Hameau is the place, a little village about ten kilos northwest of Arras. During the summer we've been pretty much all over France—Château Thierry in July, Montdidier in June, and in all the battles north of Amiens since August. I've seen every inch of the line from the sea to Rheims.

My leave came thru today, and I'm to cross Wednesday. I wish the darn leave warrant read San Francisco instead of Folkestone, but that sounds as if I'd been eating lobster again.

I had every intention of writing last night, but the colonel and the padre, he who is nicknamed "Cannibal King," came to dinner. Mess etiquette demands that flight commanders, being senior officers, should hang around and entertain. The colonel is a very decent sort, and the padre quite unlike you might expect, drinks moderately, swears mildly and very seldom, and smokes furiously. . . .

Yours completely
Bo

. . . .

Crash and grave of Captain Glentworth, the son of Earl and Lady Limerick. "To a gallant British airman" was written in German on a piece of a broken propeller.

32 Squadron,
December 2, 1918

Dearest

This is to be just a short note as I have yet to pack my few chattels and straighten the house up a bit. Tomorrow I go on leave, or at least go up to Boulogne and cross the next morning. It ought to be a decent leave, as the fellows who have just come back saw both Green and Simpson, and they want me to come and stay with them. Green lives up in Staffordshire and is on indefinite leave. His nerves were in tatters when he left here and apparently haven't recovered. Simmie is instructing just outside London, but his folks live in town [London]. . . .

Yesterday the C.O. and I went up into the awful area to look for the grave of the Viscount Glentworth who was shot down just after I came to the squadron in May. We found it and what was left of the machine and put up a cross. He was the son of Earl and Lady Limerick and a fine chap. The country is pretty awful, and there were still a couple of Huns that had been overlooked.

Today we played football against the men and came away on the short end as is usually the case.

I'm sending a few streamers and flags from London. They may look dirty and faded and undoubtedly are, but you're to take very good care of them, Izzy dear. I'll give you the details of them later.

Lots and lots of love, dear,

from Bo

13. POSTWAR *December 1918–May 1919*

ENGLISH LEAVE, FRENCH CHRISTMAS: DECEMBER 1918

The German armies retired slowly from the territories they had held in France and Belgium since 1914. On December 9, Allied troops crossed the Rhine and occupied bridgeheads agreed to in the armistice. Across the Channel, London, alight again after years of blackout, hummed with postwar activity.

. . . .

London
December 5, 1918

Dearest Izzy

... We had a nasty ride to Boulogne in a tender on Tuesday and a pretty rough ride across the Channel the next morning....

England looks very pretty now. Everything is so neat and green and peaceful. London seems to have changed a great deal already. All the lights have been cleaned of their lamp black....

The town is simply overflowing and it's almost impossible to get a room anywhere. I'm staying at the Palace Hotel, a very good place that has been taken over by the YMCA. The cafés and the theaters are full all the time, in fact it's not at all the city it was four months ago....

Yours for demobilization,
Bo

. . . .

London
December 7, 1918

Dear Lover

... Did I tell you that Tom Whitman, the last of the fellows I went to Canada with, went missing in October. I only heard a day or so ago. Of course there's a chance he may be alright. A great many people are getting back from Germany and I've already seen one or two boys I knew. They surely look funny when they first arrive, dressed up in a little of everything and most unmilitary looking. ...

Much love, lover, from
Bo

. . . .

Nile Street, Burslem,
Staffordshire
December 10, 1918

Dearest Izzy

Just another of the short notes. This time I'm waiting for Capt. W. B. Green, D.F.C., Croix de Guerre, Legion of Honor, to conclude the morning affairs of his father's business.

I came up yesterday afternoon from London to Stoke where Green met me. We had dinner with his brother-in-law and a chap named Shakespeare, no relation to Bill, in Stoke and drove out here afterwards. Wilf, meaning Capt. Green, has a nice little Darracq [French sports car], plenty of petrol, and not much to do, so I figure on a very decent few days.

The lad has received another French decoration, the Legion of Honor, and looks like a Christmas tree. He has been declared permanently unfit for flying and is on indefinite leave.

. . . .

London
December 15, 1918

... I sent you a bundle of flags and streamers a few days ago. You have them in solemn trust. If you must wash them don't wash them too clean.

The square blue and red one is the squadron flag that was over the squadron office at Montdidier, Ru[i]sseauville, Château Thierry, Bellevue, Pronville, and Douai. Obtained only after much difficulty.

The two little blue and red streamers were the C.O.'s.

The red and yellow one I carried at Château Thierry, the plain red one in the Cambrai Battle.

The Red Cross flag is a Hun one I found at our château at Douai.

The yellow rag is really a flag I found over a water depot up in front of Cambrai. It's the Sanitary Corps flag.

The silk is from a parachute flare, the parachute part.

The "officer's mess" one we had up at Bellevue.

They're all perfectly good war flags, Izzy, and you take good care of them. I've still one lovely pair of blue and red ones that I'm carrying now.

Bo

. . . .

[Izel-le-Hameau]
32 Squadron,
December 18, 1918

Dearest Id

. . . Possibly you've heard that the English Channel can be very rough on occasion. Today was one of the occasions. The boat might have been a submarine for the way it acted, and a very good prewar channel boat it was, too. Unfortunately there were a good many poor sailors aboard which didn't help matters any. We landed at Boulogne about noon and . . . just after lunch up rolled the C.O.'s car and we were back here by five. . . .

In London I met a chap who I trained with in Canada and who had just returned from Germany. He was there for over six months and had a good many interesting things to tell. His treatment was all that could be asked for, and that all the flying corps prisoners were well treated. The lucky dog is getting two months Canadian leave. Wish I'd been taken prisoner!

Yours
Bo

. . . .

December 26, 1918

Tuesday morning [December 24] we did quite a bit of flying as there is a wing aerial firing competition under way and we were picking a team. However, the morning was pretty well spoiled by a nasty accident. Four machines from another squadron had been flying a very close formation, all of them wing tip to wing tip and doing all sorts of funny things. Unfortunately one of them made a mistake, ran into another chap and cut his tail off. They were about 800 feet up, and we were all watching them.

The chap minus his tail put up a very good show and nearly got away with it. He only spun about fifty feet, broke a couple of legs, and had his head badly cut up. The other chap didn't have a hope, as both his wings fell off. It wasn't very pretty to watch. . . .

Christmas was a perfect day, at least a perfect winter day. There was just enough frozen snow to make everything pretty and not a cloud in the sky. After a church parade at ten the officers and the sergeants played football for the edification of the multitudes and incidentally for a case of champagne on the side. As usual we lost. We never can beat those sergeants.

The men had their dinner at three, and according to custom the officers and sergeants waited on them. . . .

There were the usual toasts, the King and all that, whereupon we all proceeded to the sergeants mess. They were in rare form and twas there the party began to get rough. Finally, they came over to our mess. When people started to break up the furniture, I rounded up my three sergeants and we retired to my hut and had a great old pow wow. Most of the sergeants are very decent chaps. . . . They are pretty nice people to have think well of you, for while they'll never send an unserviceable machine into the air, some machines are much better cared for than others. . . .

We had a lecture on demobilization this evening by a chap who has just taken a course on it. There are all sorts of classes and schemes, but from what I can see things look very encouraging. The whole thing is to begin about the first of February. There are several ways I can get out in a hurry, as a student with an uncompleted education, as a colonial wishing to get home at once, or as a "slip" man who has a position waiting. . . . However, I've no desire to get out before the first of February, as it will make a difference of three or four hundred dollars in gratuities. . . .

The general scheme is quite reasonable and generous, good gratuities, assistance for the men who have no jobs waiting, repatriation for everyone who came over to England from Jan. 1, 1914, education, and all sorts of things. If they will only do a bit of speeding up, everything will be lovely.

. . . .

December 28, 1918

. . . I wonder if college will really open up very much this spring? It's not too likely altho a good many fellows haven't been away long enough to make any real difference. Two years is quite a long time, besides I've other things to think about. It will be nice to see the campus as it used to be with plenty of pep and no war to worry about. It was a great old place before.

"A" flight noncommissioned officers.

. . . .

January 1, 1919

The old year wasn't a bad one when everything is considered, altho I hope never to have another just like it. We gave it a very peppy send off. . . .

Dinner was a quiet affair with only about half of the fellows here, the other half being up at Brussels. There are a good many Scotchmen in the squadron, several officers and over half of the men. For them "hog manay," as they call it, is the big day of the year. They *will* celebrate in spite of all you can do. . . . [T]he New Year was ushered in with a great deal of noise and much dancing by the Scotchmen. We came home eventually, made some rum punch, and sat in front of the fire until about four just talking.

This morning the colonel came around and wished to know why we weren't flying. Perfectly good morning it was, and not a soul in the air. So just to satisfy him two or three of us went up and looped and rolled and did all sorts of fool things above the aerodrome. It was very cold and a storm was coming, so we were only up a few minutes. . . .

Six or seven people are being demobilized in a few days, all of them men over forty-one. Only one officer is going, the equipment officer who is over the age limit, has a family and a good job at home. . . .

Bo

. . . .

[Izel-le-Hameau]
32 Squadron,
January 3, 1919

Dearest Child

... I've a grand new job coming up in a day or so, one that very few people have ever had before. Two wings in the brigade are going to hold a three days war, all aerial fighting and bombing and photography. There will be about fifteen squadrons altogether. The major, myself, and three others are to be umpires and decide who wins. The rules and plans aren't completed yet, but two of us are to judge the scraps from the air, and the rest are to observe from the ground. It ought to be a lot of fun, especially as a lot of high moguls will probably be on hand to see it. The slaughter starts in a few days. We will probably have to move up to the battlefields, which are about fifty miles square, for a day or so.

It's not everyone who can umpire a war, and I feel quite puffed up over it.

The weather—abused topic—remains awful. ...

'Night

Bo

. . . .

On His Majesty's
Service
56 Squadron,
January 7, 1919

I've been up here near Le Cateau for two days umpiring the craziest war you can imagine. The weather has been atrocious, and I had to go out with a formation this morning to see how they did their show. A bumpier, colder, windier, more cloudy, and more disagreeable hour I've never flown.

However, we are having quite a good time outside of war hours. I know quite a few fellows in 56 where the major and I are staying, and we've been visiting other squadrons every night for dinner. ...

. . . .

[Izel-le-Hameau]
32 Squadron,
January 7, 1919

The war is over and a good thing, too. I was getting fed up. This morning we had a spell of decent weather, and I went up with an early patrol from 56. They saw no "enemy" aircraft, so I came back early and went out again later. The second show was a great one. Five machines went over and straffed an "enemy" aerodrome, attacked all machines leaving the ground, and had dozens of scraps later. It was a lot of fun sitting up above and watching the show.... It hasn't been decided which side won yet....

. . . .

[Izel-le-Hameau]
January 11, 1919

Today it started raining again, a cold depressing drizzle. We had no rations issued and had bully beef for dinner, a thing that caused the mess corporal great sorrow. He maintains he has been in the squadron for two years and has never served bully for dinner, for lunch many times but never for dinner.

I'm getting active and ambitious again. Every afternoon at five I attire myself fittingly and do an hour's strenuous work at a punching bag, skip the rope, and take exercise in general. After that a bath and supper. I've been on the verge of getting disgustingly fat. Also we cut wood for our fire every morning which is real work, as the wood is always wet and tough and the axe very dull.

I've been reading John Stewart Mill, Poe, Sir Gilbert Parker, Joseph Conrad, and even old Samuel Pepys who is a most amusing old cuss as you probably are aware. Furthermore I've almost read the Bible since I came out here, something I've never had patience enough to do before....

Good night dear Isabelle,
Bo

. . . .

[Izel-le-Hameau]
32 Squadron,
January 13, 1919

Dearest

Yesterday morning the C.O. and I rode into Arras to see about our rations, which have been rather irregular of late. The trouble is with the French who have taken over the railroads in this part of the country and only permit so many trains a day for the British. That has probably been the cause of the mail being so irregular too.

Last night I went over to the sergeants mess after supper. My old flight sergeant in "A" flight went home this morning. Zink and I wanted to tell him goodbye. He's a fine chap, as honest and conscientious a man as I've

Rogers's trusted mechanics, Flight Sergeant Wellington and Sergeant Bence.

ever known, and I give him a lot for the interest he always took in my machine. He goes back to a good job with a big automobile concern. . . .

Had a letter from Shapard today, the first in a long time. He's up in Germany and pretty well fed up. The weather is wretched and the people none too congenial. Also they don't get more than enough to eat, mostly bully beef and the usual iron rations. I'm going to apply for permission to go up and see him for a day.

Love—lots of it.

Bo

. . . .

[Izel-le-Hameau]
32 Squadron,
January 16, 1919

Dearest Izzy

Truly life—expecially army life—is just one damn thing after another. Here we are settled peacefully and comfortably. Yesterday we were more or less given to understand we were to be sent home to England as a squadron very shortly, there to be demobilized. This morning we sent four machines across the channel, the first step in reducing our strength, and now someone had kicked the props out from under, taken the joy from our lives, and transformed hope and happiness into utter despair. We are to move. We are to move in a day or two. We are to move to an awful place. We are to become some comic sort of an issue squadron altho nobody knows much about it yet. We bid fair to become the hell packet squadron of the flying corps. I am on the verge of tears. What have we done to deserve a fate such as this.

This morning I nearly flew a machine back to England. It was a perfect morning, and four had to go. The thing that kept me here was the thot of a night in Boulogne on the way back. After thinking it over I'm convinced I made a mistake. It would have been very simple to have perched in the channel beside a boat, be picked up, and then wangle a decent leave on the plea of wonky nerves. Or better yet—go to Southampton, watch for a boat leaving for Canada, wait until it has gone too far to turn back, and then alight gracefully beside it. Heigh ho! I don't think very fast these days.

After getting the four machines away and cautioning the pilots not to spend more than a week in London, I had my old airship dragged out and crashed into the atmosphere, feeling madder 'n the well known March hare. Tanny went up, too, and we had a great time, broke up a formation of six machines from No. 1, had a battle with two flight commanders from

94, and then Tancock suddenly began to loop and loop and loop. I thot something had happened so that he couldn't stop, so I made a wild cross wind landing which put the wind up the Major. The aerodrome is as slippery as ice and trying to taxi cross wind you get an effect that is like a crab walking. . . .

The daily ration of love,
Bo

SERNY, THE AIRCRAFT BONEYARD: JANUARY 1919

As the delegates to the Paris Peace Conference convened for the first time on January 18, 1919, at the opulent Palace of Versailles, part of the Thirty-second Squadron moved to Serny airfield, where they would spend the next six weeks at an aircraft boneyard preparing the relics of war to be inventoried and ferried back to England.

. . . .

[Serny]
32 Squadron,
January 20, 1919

Dearest

Whatta woild!!!! Here we are in the most awful place you can imagine, everyone scattered around in four different squadrons, most of our kit still to come, weather cold and dreary, and an enormous amount of work ahead. It will take four weeks at least to finish.

We moved up here to Serny, an aerodrome just south of St. Omer on Saturday [January 18]. Having very little transport and very few men it was a pretty messy sort of a move, part of our stuff came and the rest is yet to come. We flew up in the afternoon.

Six of us are staying at a night flying squadron that is dismantling all of its machines. . . . The six are in one hut, the sort we used to have for two, with one small petrol stove. . . .

I've a ground job taking charge of all the machines that are coming in and detailing pilots to fly them over to England. I don't mind flying one over now and then but as a steady job it will get a bit monotonous.

. . . .

Rogers's "hut" at the Serny airfield with a reminder of home: on the table is a photograph of Isabelle in a frame made from a propeller tip.

[Serny]
32 Squadron,
January 23, 1919

Dearest Lover

Deliver me from the peacetime army. Oh! that the good old war were going again, that things were taking their former regular course and the only excitement an occasional tilt with the Hun. Woe—woe—woe!!

I've actually been too busy to write for two or three days. It seems as if every squadron in France were bringing their airships to us. Yesterday morning I was eating breakfast and indulging in a bit of self congratulation upon having arisen so early, when the sky simply began to rain aeroplanes. They all have to be checked up, vouchers signed and a lot more fool stuff. They have finally been put safely away in the hangars, the log books must be made up and signed and reports signed and sent away and all sorts of red tape. We've taken none over to England yet and have quite a collection of them on hand. There are more coming tomorrow and even more the day after. . . .

Just across the road there is an aerodrome where many Fokker Biplanes and other Hun machines are stored. The C.O. promised to get me one to play with.

Golly, Izzy dear, it will be a large day when the king has had enough of my services and presents me with a ticket home. Lots and lots of love,
Bo

Rogers's favorite SE-5A, number D6996.

. . . .

[Serny]
32 Squadron,
January 25, 1919

Dearest Izzy

... The old SE 5s are pretty fine airships and the day mine has to be handed over I know I'll weep copious tears. It's like getting rid of a horse you've had for years. Aeroplanes have personalities. When you've had one for months and months, when it's always brought you home safely and never let you down in a pinch you acquire a small bit of affection for it. I'd like to be able to bring mine home and store it in the barn.

Good night, dear Isabelle.

Bo

CROSS-CHANNEL FERRY FLIGHTS: FEBRUARY 1919

Two and a half months had passed since the end of hostilities. Millions of Americans and Canadians strove to get a berth on a ship sailing for North American. Despite bad weather, ferry flights continued through February. By the end of the month, only five machines remained at Serny. Major Russell was reassigned, and Captain Rogers was left in command of the Thirty-second Squadron.

. . . .

Savoy Hotel, London
January 28, 1919

Dear Child

This afternoon I took my first little flip across the channel. It's an over-rated pastime. I didn't enjoy a minute of it until it was all over, then it seemed quite amusing.

When we left Serny this morning about eleven it was clear. I had four fellows with me, and we all landed at Marquise, an aerodrome near Boulogne.

We left there about one, climbed up to 5000 before starting across, and from then on saw nothing but a solid bank of clouds below us. There was a strong drift wind, so I set a compass course and tried to navigate. After half an hour I figured Folkestone should have been below, but for all I knew it might have been the North Sea or the Atlantic. Finally I went thru the clouds, a solid mass of them 4000 feet deep. About the time we should have hit the ground, it suddenly became clear, an expanse of snow covered fields only 500 feet below. Believe me, it was the most beautiful sight I've ever seen. Had there been water instead of land I'd have given up the ghost.

Having not the remotest idea where I was and having lost the other four, I landed in a field where an old man told me the nearby village was Cranbrook, miles from where I should have been. After nearly hurling a death on a hedge almost too high to clear I got to the right aerodrome.

Fifteen of us left Serny. Eight got over here. Goodness knows where the rest are. My four must be O.K. for they went down over land.

Them as wants can attempt to fly the Atlantic. I don't need the £10,000 badly enough to try after this afternoon's experience. . . .

Good night lover dear,
Bo

. . . .

[Serny]
32 Squadron,
January 31, 1919

Dearest Izzy

. . . Wednesday morning I went down to Folkestone on the staff train and had lunch with Major Young, the C.O. of Number 1 Squadron. We crossed over in the afternoon, but as our transport was mixed up, I had to stay in Boulogne over night. Deadly hole! There I met a chap who told me what happened to Tom Whitman. He was hit by one of our own shells and fell to pieces some distance up. . . .

Last night we went down to a nearby village for dinner. . . . When we came home we found several fellows just back. They all had wild tales to tell. Some of them landed miles from where they should have been, and all of them agreed it was an unpleasant trip. One chap ran into a tree while hedge hopping, crashed his machine quite badly, blackened both of his eyes, and broke a perfectly good camera that was in his pocket. Such is life.

My Fokker is coming tomorrow. A very nice machine it is but painted quite plainly.

Much love from

Bo

. . . .

> The Grand,
> Folkestone
> February 8, 1919

Dearest

This ferrying job is getting to be quite an international occupation. One day you're in France, the next day England.

Four of us came over this afternoon, and this crossing was as pleasant as the last was disagreeable. France and England are covered with snow, just one great glistening patchwork. We left Serny shortly after lunch, landed at Marquise long enough to report and get life belts, and came directly to Lympne. We had no more than gotten off the ground at Marquise than we could see the cliffs at Dover and the snow covered hills behind. We crossed at 4000 feet, the sea a shimmering golden carpet and the sky a misty blue. The south coast was visible for miles, Dover, Folkestone, Hastings, Beachy Head, and dimly in the distance, Brighton. To the north and east we could see Ramsgate and Margate and even the Thames Estuary. There were boats everywhere, some of them apparently stationary, glued to the carpet, others leaving long white wakes behind. It was the sort of day you feel like looping and rolling and spinning and playing the fool generally. Unfortunately one of my landing wires [bracing wire between wings] broke on the way across, so my desires were unsatisfied.

. . . We are crossing back in the morning. . . .

Much, much love, dear.

Bo

. . . .

A Fokker D-7, probably at the Serney airfield, 1919.

[Serny]
February 12, 1919

... The Fokker came this afternoon after much delay, but sounded as if it were suffering from a touch of flu. It sputtered and popped and sounded quite unhealthy. Aside from that it's a rather nice airship. ...

. . . .

[Serny]
February 14, 1919

Nine more machines away today, which nearly finishes the lot, but more to come shortly. The weather is turning dud, a few drops of rain tonight and many clouds. No mail. The Fokker is busted. A valve went yesterday, so we'll have to send it back on a lorry. I've been getting clubby with the wing doctor and expect—aye, intend—to have a nervous breakdown as soon as our job is done here, go to 14 General Hospital in Boulogne and wangle leave or repatriation. It's a lovely hospital and the C.O. is very partial to flying corps people.

I've been out in this Godforsaken country long enough and intend to get out of it. ...

. . . .

[Serny]
February 18, 1919

Indeed, your old lover is a most sad and dejected youth this evening. A vast sorrow oppresses him, crushes his spirit and stifles his hopes. Woe! Woe! Woe! 'Tis my lot to be cursed with perfect health, heart normal (physically), pulse strong and normal, nerves none at all, in fact everything present and correct, sir. . . .

This morning I went to see the wing doctor. Says I, "Doc, I've been out here ten long months, have done many hours flying and should according to Hoyle, be a sick man."

"Very good," says he, "We shall see." And he takes out a form, puts down name, rank, age, time in France, hours flown, how many cigarettes a day, and a few more items. Then he tested me, hammered my chest, listened at the heart thru that little phonograph arrangement, had me standing on one foot with eyes shut, hammered my funny bone, pounded my knees, put me thru a course of Swedish exercises, and then felt my pulse. After my garments were rearranged he asked, "Do you sleep well? Eat well? Dream? Are you keen on flying? Have you a hectic cough? etc." After telling him I slept like a log, was possessed of a mighty appetite, flew at every opportunity, coughed only while smoking issue cigarettes, and denied dreaming, which was a shameful lie for I dream of you, not often enough, but still often, my dear, he said I was a perfect 36, or words to that effect, but for friendship's sake he'd see what could be done as I'd been out here much longer than was necessary.

All of which is very indefinite, but which may turn out advantageously. Anyway, demobilization, or I should say, repatriation of Canadians is coming along very nicely. I live in hope. . . .

. . . .

[Serny]
February 23, 1919

For a week or so I've been acting adjutant. It's not the easy job I'd always imagined it to be. We are demobilizing about ten men a day, mostly from other squadrons, and the papers for each man require some ten signatures. There are dozens of odd little details to be attended to every day. I'll be glad when we get a regular adjutant. . . .

. . . .

[Serny]
February 27, 1919

... The cadre, which you may recognize as French for framework, of the squadron is going back to England on the third of March. Veitch and Tancock take it home. All the squadron records and fourteen men.

The major went up to Cologne today to take over a squadron there, and I am in command of the squadron, a real job, as there is a lot of work to be done these last few days. This evening the colonel called up about several things, ticked me off for not having sent a party out to a machine that had a forced landing near Hazebrouck, wanted a man for the wing, and spoke of some dozen other small details. To make matters worse there is only one old man left in the office. All officers and men are to be posted away in a day or two. ...

We hurled a hefty party last evening in honor of the major. There were only a few here as most of the fellows are still in England, but we made up in enthusiasm what we lacked in numbers. ...

I hated to see the major go, for he's a splendid chap and has certainly treated me white. He's taking over one of the best squadrons in France and will probably get to India in a few months. Incidentally he told me just before leaving that Vietch and I had been put in for two decorations, one the Distinguished Flying Cross and the other a comic Roumanian medal [Order of Michael the Brave]. Being recommended doesn't mean particularly that will get them, altho the Roumanian one is practically sure, as it's an issue affair.

. . . .

32 Squadron RAF,
March 2, 1919

Dearest Isabelle,

I'm head over heals in work. Veitch hasn't come back from England yet. If he doesn't show up tonight, I'm taking the squadron to England. If he does come back, I'm going down to Boulogne with the cadre and then go to England via 14 General Hospital. The doc gave me all the necessary papers this morning.

There surely are tons of things to be done at the last minute and nobody to do them. All the good men have been demobilized, and the riffraff don't seem able to carry on. ...

Bogart Rogers, Capt.
Commanding No. 32 Squadron

Major J. C. Russell, commanding officer of the Thirty-second Squadron.

RAF
In the Field,
2nd March, 1919

THE THIRTY-SECOND SQUADRON RETURNS TO ENGLAND:
MARCH 1919

The cadre of the Thirty-second Squadron, consisting of fourteen men and all the records, settled at Tangmere. The idle squadrons were available for such a duty as flying the mail, as their men awaited demobilization.

. . . .

American YMCA
London
March 5, 1919

Dearest Lover

France is behind. Dirty, smelly, disagreeable, depressing old Boulogne is on the other side of the misty, cold rough old Channel. May they both roast in hell, a mixed metaphor.

I came over yesterday with the cadre, not in charge for Veitch came back, but with it nevertheless and posted to Home establishment for a rest. Tomorrow morning I sneak up to the Air Ministry and get some dope on the future. From what I can find out the quickest way home is to get your own passage. This repatriation business is infernally slow. . . .

The cadre has gone to Tangmere, the place where I trained last spring. I may go down there and stay for a week or I may go up to Green's, possibly both. Much depends upon what I find out tomorrow.

Love from,
Bo

. . . .

American YMCA
London
March 7, 1919

. . . I'm disappointed, disillusioned, and a wee bit discouraged. Most of the day I've spent down in the City trying to find out about boats and sailings. I always imagined that from England one could get a boat to almost any part of the world but 'tis not so. Apparently there are none from here to the coast via Panama. There are plenty to Canada and the states. The White Star people said they might possibly book me passage for the middle of April, but it wasn't too certain.

The very bright idea of finding a good freighter was knocked in the head when I found that the Ministry of shipping won't permit passengers on freighters.

Yesterday I reported to the Air Board, got two weeks leave, and at the end of that am to be posted somewhere in the southeastern area, probably Northolt as Simpson is trying to get me a flight out there. I went to see Simmie last evening, had dinner and came back in the evening. . . .

There are dozens of fellows hanging around waiting to get home, all of them fed up completely, all of them without interest, pep, or ambition, all wanting merely to get home. . . .

. . . .

God Begot House,
Winchester
March 11, 1919

. . . At Winchester, after bearding the lions in their dens, I found that my official status is as follows: i.e. and to wit, passage has been granted. All that remains to be done is wait until there is a vacancy at the repatriation camp at Winchester, then wait there for a boat. Not much of an outlook but yet a small amount of progress in the right direction.

There have been one or two rather bad outbreaks by Canadians and Americans waiting to get home. It may wake the powers who be. . . .

. . . .

Savoy Hotel
London
March 13, 1919

Who should I bump into this morning in the Strand but Shapard who has just come over from France. Like so many others he is awaiting repatriation. They had him down at Blamford, in Dorset, an awful camp, but he wangled a two weeks' leave. I had perfectly good intentions of going down to Tangmere this afternoon, but Shap said he might go along if I waited until tomorrow. The chance that he may warrants the wait. . . .

. . . .

32 Squadron,
Tangmere, Sussex
March 16, 1919

. . . Yesterday morning in town [London] as I was strolling into the Chancellor of the Exchecquer's—Mr. Cox's well-known bank—I ran into Veitch who had just arrived in town . . . and was on his way to Herne Bay or some such place. Therefore I took an afternoon train from Victoria to Chichester, found a tender coming out here to Tangmere, and am now sleeping in Veitch's bed. . . .

The place has changed since I was here this time last year. Then it was only about half finished, with mud everywhere and most uncomfortable. Now it is a lovely permanent station with paved roads running all over the place, fine permanent quarters and hangars, tennis courts, and all sorts of comforts that one has no right to expect in the Army. . . .

. . . .

Northolt,
March 21, 1919

Thank goodness . . . I'm settled safely and comfortably at Northolt. Yesterday morning I reported for duty, was asked where I wanted to go, and chose this station in preference to any others I could have gone to.

It's quite a fair place, only half an hour from town [London], Simmie is here, and there are several other people I've known at various places. There's not much work and very little flying. . . .

I'm so completely fed up with this damned little island and the impossible weather that it makes very little difference one way or the other.

. . . .

Hounslow,
March 25, 1919

Life almost promised to pep up a bit yesterday and to furnish a dash of excitement but tonight it seems to be different. It all started by several machines and pilots being sent from Northolt over here to Hounslow for temporary duty. Nobody had the least idea what it was all about but Simmie said he'd take a machine if I would. We came over expecting to come back the same afternoon. Upon arriving here we found that our job had something to do with the impending railway strike. There might be some excitement, and we were under no circumstances to leave the aerodrome without special permission. How long we are to remain nobody seems to know. . . .

December 1918–May 1919

Nothing has happened today, and it's doubtful if anything will. However, it's a change and helps pass the time. Anything that speeds the time is welcome. If the strike should really take place, which is doubtful now, we might have an exciting job to do.

. . . .

Hounslow,
March 27, 1919

Yesterday we received secret and solemn orders, which are not to be told even to other people doing the same job. Simmie and I are working together, he as pilot and myself as his observer. We are to fly in Mons Avros, very peaceful old school machines. We were going out to rehearse our show this morning, but it was entirely too windy for any sort of flying. . . .

. . . .

Hounslow,
March 31, 1919

Dearest Child

Yesterday morning Simmie and I started away to do our job, a 300 mile cross country affair with several landings and some red tape in connection with said landings. It was a perfect morning when we started, not a cloud in the sky.

We landed first at Northolt to get the mail, and then went over London and up into Essex where we were to make the second landing. It was horribly bumpy, and we were bounced all over the place. Eventually we went down to the tree tops where it wasn't so bad, but much harder to follow our course. Simmie did the flying while I fussed with maps. After the second landing we headed off to Norfolk and arrived at the end of the route about noon. It was snowing and blowing a gale when we perched.

After reporting we went up to the mess for lunch, came back to see that the machine was filled with petrol and oil, and waited for the storm to let up a bit. We finally got away and for nearly two hours fought against the wind, flew thru storms and had a generally disagreeable time. We had to land for petrol about half way home and eventually got back here to Hounslow about five thirty. It would have been a nice trip if it hadn't been for the wind and snow and the awful bumps. The country we flew over is as flat as a table. In one harbor on the coast we could see a large part of the navy and a few German ships.

What will happen when we get back to Northolt is a mystery. They are repatriating and demobilizing everyone there so I have hopes. Thank goodness one can always hope.

A great deal of love from,
Bo

THE WAIT FOR REPATRIATION: APRIL–MAY 1919

Shiploads of Americans and Canadians were sailing every day from Liverpool. In March, the RAF inaugurated a regular cross-Channel airmail service from Folkestone to Cologne to support the British Army of Occupation. But the most-talked-about event of April 1919 was the competition for a transatlantic flight spurred on by a £10,000 prize offered by the London *Daily Mail*.

. . . .

Hounslow,
April 3, 1919

Dearest

Yesterday Simmie and I did a real job of work, in fact everyone did. We received regular orders just before lunch, such affairs always happen just at meal time, and pushed off immediately. Our trip was the same one we did last Sunday, only we had to do it in much better time. Up along the east coast the weather was wonderful. We enjoyed every minute of it.

On the way back Simmie decided to sleep, so I flew most of the way. Now and then I'd do something funny to wake him up, whereupon he would shut off the engine crawl over the back of his seat and make a few rude remarks.

There's nothing to beat flying when the weather is right and you've got a congenial passenger and a machine that gets where you want to go. I don't care for rough days, low clouds, or mist, but when the sun shines and there's a nice steady wind—. . . .

Yours
Bo

. . . .

Royal Air Force Club
Bruton Street, W.1.
April 5, 1919

Dearest Lover

. . . I've been having a great time lately flying in the back seat of a lovely Bristol Fighter belonging to one Capt. Bill Pace who uses said machine for dispatch work. Pace, by the way, is the man who led the two daylight bombing shows to Cologne. He has some wonderful pictures of the raids.

Yesterday morning Simmie and I squeezed into the back seat and were taken over to Northolt where we intended to get the mail. However, it had been forwarded to Hounslow and hasn't arrived there yet. . . .

Love from
Bo

. . . .

Hounslow,
April 7, 1919

Dearest

. . . In the afternoon Simmie, Pace, and I played golf, if banging the small ball around the aerodrome may be dignified by that name. The weather had been quite respectable and I feel like a human being for a change.

Pace pushed off for Paris this morning in a Martinsyde, trying for the record there and return. Unfortunately he had engine trouble coming back and is in France somewhere. He is going to Madrid in a week or so. If I'm still around I'm trying to arrange to go with him. However, there's not much hope, as I don't belong to the station or the squadron. . . .

The demobilization officer at Northolt told me yesterday that some more people should be leaving there very soon. However one becomes a bit pessimistic about anything that is said in regard to demobilization, repatriation, and the like.

'Night
Yours,
Bo

. . . .

Piccadilly Hotel
London,
April 11, 1919

Dearest Id

. . . Simmie most unexpectedly got orders to Cologne. He leaves tomorrow to take a flight or possibly a squadron. He was shaken up over it as he expected orders to Egypt in a month or so. Naturally I feel pretty much lost now that he's going away.

A few day ago I had a lovely ride with Bill Pace in his own D.H.4. We went over London, then down the river and had a great time playing around in the clouds. As Bill wanted to make a few tests we got up to 10,000 and the old bus gave 106 m.p.h. full out which is pretty fast for that height. It is the best machine I've been in from a passenger's point of view. The back seat is roomy and comfortable, and not a breath of wind gets in anywhere. The only drawback is that the pilot and passenger are separated by a large petrol tank and can only talk by telephones.

A great deal of love from,
Bo

. . . .

                                                Hounslow,
                                           April 13, 1919

Dearest Child

Simmie and I had a very quiet and somewhat gloomy party last evening. . . .

This morning Bill Pace took me over to Hendon in his Bristol Fighter. It was clear and warm and the machine is a lovely affair. Bill has applied for service overseas, France, Russia, Egypt, anyplace to get away from England. He has the right idea for England isn't much of a place.

Yours
Bo

. . . .

                                                Hounslow,
                                           April 14, 1919

Dearest

. . . Yesterday Pace and I almost went to Paris. He was to take some dispatches over, and I was booked as Passenger, but at the last minute the show was called off. It would have been a nice trip. Last week a chap from here went over in seventy-five minutes. It's a matter of well over 200 miles.

Love from,
Bo

. . . .

                                                Hounslow,
                                           April 18, 1919

After an important conference with Mr. Winston Churchill this morning I'm seriously considering not speaking or writing to anyone—that is, to none of the proletariat. Of course you aren't to be included under that heading and yet—well, the Secretary of State for War is just that. As I said before it was a most important conference. Possibly you might be interested in hearing the conversation verbatim.

The transcript:

Capt. Bill Pace and Capt. B. Rogers—in unison—"Good morning, sir."

Mr. Winston Churchill, Secretary of State for War—"Good morning. Did you have any trouble finding the way?"

Capt. Bill P.—"No, sir."

* * *

Denoting lapse of time.

Mr. W. C.—S. of S. for W. "Are you pushing off[?]"

Capt. B. P. and Capt. B. R.—in chorus and very expectantly as it was nearly lunch time and Winnie's Rolls was standing by—"Yes, sir."

Mr. W. C.—S. of S. for W.—delivering the blow—"Good bye and many thanks for your trouble."

Capts. B. P. and B. R.—as one—"Good bye, sir."

And so saying we crashed into the void and came back to Hounslow.

As I see you are a bit puzzled permit me to clear up the situation with a few details.

A chap here, Capt. Lloyd, is teaching Mr. Churchill to fly. This weekend being a holiday [Easter] Mr. C. decided to take his lessons at home and phoned for Lloyd to come down to said home in Kent. Lloyd went down in an Avro with a mechanic in the front seat. Bill and I followed in the Bristol with some flying kit and several tins of petrol. On the way home we dived on an elegant golf course and scattered old men, caddies, clubs, balls, and green flags all over the place. No doubt it put some of the old boys off their game a bit. . . .

Had a letter from Major Russell this morning saying among other things that my D.F.C. would be either in the birthday honors or the peace honors, and that I might wear the riband.

. . . .

Hounslow,
Easter Sunday
[April 20, 1919]

Yesterday we flew a bit in the morning and went over to Twickenham in the afternoon to see France and New Zealand play rugby. . . .

The game was not much from a rugby point of view—New Zealand won, 20 to 3—but the crowd was large and distinguished. The king was present with the four princes, all of them being presented to the players—

or vice versa. Then there was Sir Douglas Haig and dozens of other generals, admirals, and other notable and noble personages.

The outstanding feature of the play (perhaps it should not be mentioned had it not caused the king, the four princes, Sir Duggie, and practically all the other men to laugh heartily and the ladies to laugh mildly but blush deeply) was a son of France, in the heat of the battle, having his—er-r—his nether garments torn from his person. No sooner the foul deed done than the play and players moved away leaving him alone before the multitudes, looking wildly but in vain for a hiding place. After a short (but doubtless embarrassing) interval he discovered the damaged garment not far away, made a dash for it, girded it gracefully about his loins, and the day was saved for France.

. . . .

Hounslow,
April 23, 1919

There is a *possibility* that something has been wangled. Before many more moons somebody may see about a ticket home and a seat on a boat. I'll feel quite sure about it when on the bounding main and not before. Anyway I've caused a bit of bother here and there. Maybe something will come of it. You never know.

. . . .

Hounslow,
April 24, 1919

Dearest

Your letters make me want to get home worse than ever, if such a thing is possible. However it may not be necessary to write very many more letters. In the morning something definite may come thru. . . .

I've been getting some civilian clothes. They have lovely cloth over here, tweeds and such, but somehow or other the boys don't seem to know how to sew them together decently. I'm going to try getting some cloth made into rugs or blankets so I can get it home duty free and have it made up there.

Bill Pace is going to Madrid in a few days on some sort of a show. After a short stay there he is doing a tour of France. Providing he has no bad luck, it should be a wonderful trip.*

---

*Tragically, Captain Bill Pace died in an automobile accident a week later, on May 2, 1919.

Now that I come to think of it, today is the anniversary of my going to France. A year ago tonight I was on my way from Boulogne to the pool at Berck. It doesn't seem so very long, but it's been a most eventful year. Wonder where I'll be this time next year?

Yours,
Bo

Rogers received his orders home in the next day or so and decided to write no more letters because he expected to beat them home. He telephoned Isabelle from New York City, spent a few days with family friends, and caught a train for the west coast. A Western Union Telegram dated May 21, 1919, from Ogden, Utah, read:

TOMORROW NIGHT'S THE NIGHT.

POSTSCRIPT

Bogart Rogers arrived in San Francisco on May 22, 1919, borrowed a car, and drove to Palo Alto, where, for the first time in almost two years, he and Isabelle were reunited on the steps of the Kappa Alpha Theta sorority house.

They were married on June 26, 1920, at Isabelle's home in Albany, Oregon—she dazzling in white silk and he splendid in his RAF uniform. The newlyweds moved to southern California and rented a house in Hollywood.

Bogart never did return to Stanford, nor did he follow in his father's footsteps. He had found a flying job. Employed by the *Los Angeles Examiner*, he flew photographers to newsworthy events and rushed the photographs back in time for the late edition.

Around 1925, Bogart and Isabelle bought a small two-bedroom house in Beverly Hills, a new community just west of Hollywood. They would live there and raise a family over the next twenty-five years.

Following a brief newspaper career, Bogart was hired by the Hal Roach Studios as the business manager for silent-film actor Douglas Maclean. At the same time, he began developing a talent for fiction writing and was soon selling short stories and articles about flying to *Cosmopolitan*, *Red Book*, *Popular Aviation*, and *Liberty*.

The movie industry of the 1920s had become a magnet of sorts for former combat pilots. "Wild Bill" Wellman directed *Wings*, a film that starred Brooklyn-born Clara Bow, a seductive auburn-haired beauty whose business affairs Bogart also managed for a short time, though his interests lay elsewhere.

In February 1929, the city of Glendale, California, held the opening ceremony for Grand Central Air Terminal, the most auspicious airport yet built to serve the greater Los Angeles area. Following the dedication,

Bogart was appointed as Grand Central's first general manager, a post he held long enough to witness the inauguration of coast-to-coast airline service.

By 1933, he had written and produced his first motion picture for Paramount: *The Eagle and the Hawk*, about two American fliers in France, starred Fredric March, Cary Grant, and Carole Lombard. In 1936, he produced *Pigskin Parade*, a movie with Judy Garland and Jack Haley about a country farmer who becomes a college football hero.

Near the end of the 1930s, he invested in a horse-race photo-finish camera, an invention that, to his astonishment, turned out to be so successful that he formed a corporation, Photochart, Inc., and became its president. The thoroughbred horse-racing industry opened up a new opportunity for writing.

After the United States entered World War II, Bogart's friend Eddie Rickenbacker, president of Eastern Airlines, offered to help him obtain a commission in the Army Air Corps.

Years had passed since he had held the controls of an airplane, and he may not have been happy with a desk job, so he chose instead to write about the war in numerous magazine articles and a book about the Flying Tigers.

Both his younger brothers had joined the Marine Corps. In 1944, the older of the two, Thornwell, a young lawyer before the war, disappeared while on a bombing mission to Rabaul. His death scarred Bogart deeply and permanently. The youngest brother, Bryson, survived the savage fighting on Iwo Jima.

He was a family man. In his own words (circa 1945),

> I have a charming wife who, I fear, understands me; a son whose misfortune it is to look like me; and a very snazzy daughter who writes from Stanford, where she is a freshman, that she loves me and if I should see anything in Fifth Avenue shop windows that I think she might like, she's quite sure that she would.

A man of great personal charm, he made friends easily and never gave up the habit of writing letters. Correspondence was his way of keeping in touch with his many friends around the country and abroad. A wide range of interests kept his mind always working on a new story idea. With the war ended and the racetracks reopened, Photochart prospered. His writing took a new turn. Horse-racing and sports in general became the source of his stories. With his agent and editors in New York and Photochart in California, he was traveling a great deal between two coasts.

In December 1960, he attended a reunion of World War I pilots in

New York. The Air Force Museum captured it on film. If this were a historic moment, the subjects, busy swapping war stories, were unaware of its import. A larger reunion was scheduled for June 1961, but Bogart did not attend that one.

He was hospitalized with a stroke, suffering partial paralysis of his right side. Eventually he recovered, but he was not writing as much. Managing the photo-finish activities for thirty-five horse-racing associations as president of Photochart took most of his time. It was during the 1966 racing season that he collapsed with a blood clot in his leg at the Hollywood Park racetrack. He was taken to St. Joseph's Hospital in Burbank, where he died the next day, July 24, one month after his sixty-ninth birthday and forty-eight years since he had fought in the Great War.

INDEX

Albany, 123, 152, 156
Albatross, 125, 127, 128
Alcock, Cadet (killed in training), 37
Archie, explained, 107
Armistice day letter, 215
Austro-Hungarian peace note, 196
Avro, 62, 85
Ayr, 88–94

Bairnsfather, Bruce, 84
Bapaume, battle of, 165
Beauvois, 100–123
Belasco, David, 50
Bellevue, 162–208
Bellicourt, 171
Berck, 94–100
Bishop, Major William, 23, 134
Boulogne, 94, 161, 229, 247
Bristol Fighters, 193
Bruges, 205
Bulgaria, surrender of, 194, 196

Calgarian, 70, 72
Calhoun, Ellen, 23, 46, 52, 69
Callender, Lieutenant Alvin, 125, 156, 197, 198
 grave of, 218, 223, 224
 killed, 213, 216
Cambrai, 171, 194
Camp Benbrook, 36
Camp Borden, 17, 26, 30, 31, 35

Camp Bowie, 38
 quarantined, 37
Camp Everman, 36
Camp Hicks, 41
Camp Mohawk, 17, 27, 28
Castle, Captain Vernon, 35
Château Thierry, 157
Chattis Hill. *See* Stockbridge
Chinese coolies, 53
Churchill, Winston, 253
Claydon, Captain Arthur, 128, 135
Cohan, George, 50
Coldstream Guards Band, 189
Corbett, Jim, 50
Corse, Cadet, 35, 187
Cox & Co., 74, 154, 155, 160, 205
Crabb, 168, 170
Crary, Allan, 150, 161
Crary, George, 206
Cupelle, Captain, 41
Curtiss Jenny, 17, 62

De Palma, Ralph, 50
Deseronto, 17, 27, 29, 30, 31
Dolphin scout, 23
Donaldson, Lieutenant J. O., 220
Douai, 171, 194, 202

Eastman, Joe, 152

51st Division, 189
Fisher, Harrison, 50

Flanders, 194
Flu epidemic, 214, 226
Flynn, Jerry, 107, 140, 142
  killed, 176
  made captain, 144
Fokker, 195
Fokker Biplane, 137, 139, 149, 151, 172, 239
Fokker Triplane, 123, 139, 142
Fonck, Rene, 151
Fouquerolles, 124—131
French, General John Denton, 64

Gard, Deke, 214
Gibbons (Reckless Reggie), 67, 70
Gibson, Bill, 150
Glentworth, Captain (Viscount),
  grave found, 227
Gorgas, General William C., 33, 42
Graham, Lieutenant Robert, 126
Green, Captain Wilfred B., 131, 142, 151, 166, 230
  avenges Flynn's death, 178
  awarded Croix de Guerre, 162
  awarded DFC, 187

Haig, General Douglas, 255
Hale, Frank Lucien (Bud), 180
Halifax, ruined by explosion, 53
Hearn, Lieutenant, 37, 41
Hendrie, Scottie, 107
Herriman, George (Krazy Kat cartoonist), 50
Hounslow, 156, 158, 249

Immelman turn, 83
Imo, Belgian relief boat, 54
Izel le Hameau, 219—228, 231—238

Jackson, H. H. (Harry), 33, 41, 44, 76, 81, 156, 158
  missing, 168
  posted to 28th U.S. Squadron, 45
Jarvis, Lieutenant E. M., 115
Jeffers, Johnny, 151, 152

Jellicoe, Admiral John Rushworth, 64
Justicia, 152

Khaki Club, 26
Koerner, Dutch, 214

La Brayelle, 211—218
Lawson, Bruce, 187, 191, 220
Leaf, Bill, 134, 154, 156, 158, 161
  killed, 216
Le Cateau, 234
Le Catelet, 171
Leland Stanford Junior University
  (L.S.J.U.), 23, 39. *See also*
  Stanford
Lewis gun, 129
Liberty Motor, 50
Liege, 194
Lille, 194, 202
  captured, 205
Lloyd, Captain, teaching Churchill to fly, 254
London, 154—161
Lympne, 242
Lys, 194
Lytle, Bob, 161, 190, 195, 224

McBean, 220
McCudden, Captain James, 90, 91
McMullen, Captain, 23
Major, the. *See* Russell, Major John C.
Marquise, 241, 242
Marriage proposal, Isabelle accepts, 46
Mestayer, Harry, 50
Moncton, site of German prison camp, 52
Morocco, Walter, 50

Namur, 194
Naval Air Service, 26
Ninth Brigade, RAF, 122, 138, 171
Northolt, 158, 249

Ostend, 202
  captured, 205

Pace, Captain Bill, 251, 252, 253, 254
  killed in auto accident, 255
Palo Alto, 20, 66, 190
Paris trip, 150–151
Paskill, Lieutenant Reuben Lee, 133,
    145, 150, 156
  killed, 162
Peace propaganda, 196
Pickford, Jack, 50
Pronville, 208–210

Reagan, Bill, 151, 155
Reckless Reggie (Gibbons), 67, 70
Rogan, Major MacKenzie, leader of
    Coldstream Guards Band, 189,
    190, 196
Rogers, Cadet Bogart
  posted to Camp Benbrook, 36
  posted to Camp Borden, 30
  posted to Camp Hicks, 41
Rogers, Captain Bogart
  award of D.F.C. confirmed, 254
  cross channel ferry flight, 240, 242
  describes battles engaged in, 226
  describes war flags sent to Isabelle,
    230–231
  does job for Churchill, 254
  promoted to captain, 225
  recommended for two medals, 245
  reports for duty at Northolt, 249
  temporary duty at Hounslow, 249
  waits for repatriation, 251
Rogers, Lieutenant Bogart
  deputy flight leader, 136
  double victory, 179
  eyewitness to sinking of *Tuscania*,
    60–61
  first assigned to 32nd Squadron,
    RAF, 100
  first flight in SE5A, 79
  first victory, 151
  last victory, 212
  posted to Ayr, Scotland, 88
  posted to Chattis Hill aerodrome,
    64

  posted to Tangmere, 75
  promoted to 1st lieutenant, 85
  single victory, 185, 191
Rotary engine, described, 73
Ruisseauville, 132–145
Russell, Major John C., C.O. of 32nd
    Squadron, 124, 132, 133, 144, 155,
    177, 180, 187, 197, 207, 214, 224,
    225, 227, 234, 245, 254

St. Johns, Adela Rogers, 130
Saint-Quentin, 171
Sandys-Winsch, A. H., 176
School of Military Aeronautics No. 4,
    17, 20
Seagrove, Gordon, 48
SE-5, 99
SE-5A, 62, 79
Serny airfield, 238–245
Shapard, Evander (Shap), 64, 65, 66,
    82, 88, 89, 91, 168, 170, 183, 193,
    197, 199, 205, 248
  letter of, 184
Simpson, Captain S. P. (Simmie), 107,
    116, 117, 135, 140, 142, 151, 167, 187,
    250, 251
  awarded Croix de Guerre, 162
  departs for Cologne, 252
  returns to England, 163
Smith, Andy, fraternity brother, 26,
    222
Smith, Dave, fraternity brother, 204
Sopwith Camel, 62
Sopwith Pup, 22
Squadron
  32nd RAF, 100, 247
  56th RAF, 234
  78th Training, 36
  79th Training, 36
  81st Training, 36
  83rd Training, 36
  87th Training, 36
  91st Training, RFC, 75, 77
  92nd Training, RFC, 75
  93rd Training, RFC, 77

Stanford, 20, 27, 38, 42, 145, 148, 151, 214. *See also* Leland Stanford Junior University
Stockbridge, 64–75

Tancock, Monte, 173, 186, 195, 199, 202, 215, 245
Tangmere, 75–87, 249
Taylor, Bill, 22, 24, 36, 37, 50, 52, 93, 99
  confirmed killed, 168
  reported missing, 134
Taylor, Palmer (Pop), 19, 23, 25, 31, 36, 52, 69
  dies of scarlet fever, 46
Thomas, Lieutenant C. W. (Tommy), 186
Thompson, Major, 192, 196
Touquin, 145–153
Triplane. *See* Fokker Triplane
Trusler, John, 177, 203
Tunisian, 53
Turkey, 196
Tuscania, sinking of, 58–59, 60–61

Tyrrell, Captain Walter A. (Bing), 107, 126

Veitch, Captain, 224, 225, 245, 249
Vickers Machine Gun, 23

Wheatly, Winnie, fraternity brother, 26
Whitman, Tom, 134, 156
  death decribed, 242
  missing, 230
Williams, Captain, 41
Wilson, Padre (Cannibal King), 182, 226
Wing
  Elementary Training, 30
  42nd Training, 30, 36
  43rd Training, 36

Young, Major, C.O. of No. 1 Squadron, 241
Ypres, 194

Zacharias, George, 151
Zink, Captain, 167, 172, 184, 195, 197